Ergonomics Tool Kit

Practical Applications

Mike Burke, PT
Ergonomics Consultant
Ergonomix, Inc.
Lake Zurich, Illinois

AN ASPEN PUBLICATION®
Aspen Publishers, Inc.
Gaithersburg, Maryland
1998

Library of Congress Cataloging-in-Publication Data

Burke, Mike
  Ergonomics tool kit: practical applications/
    Mike Burke.
       p.   cm.
    Includes index.
    ISBN 0-8342-1006-1
    1. Human engineering. I. Title.
    TA166.B893   1997
    613.6'2—dc21
                                97-36248
                                    CIP

Copyright © 1998 by Aspen Publishers, Inc.
All rights reserved.

Aspen Publishers, Inc., grants permission for photocopying for limited personal or internal use. This consent does not extend to other kinds of copying, such as copying for general distribution, for advertising or promotional purposes, for creating new collective works, or for resale. For information, address Aspen Publishers, Inc., Permissions Department, 200 Orchard Ridge Drive, Suite 200, Gaithersburg, Maryland 20878.

Orders: (800) 638-8437
Customer Service: (800) 234-1660

About Aspen Publishers • For more than 35 years, Aspen has been a leading professional publisher in a variety of disciplines. Aspen's vast information resources are available in both print and electronic formats. We are committed to providing the highest quality information available in the most appropriate format for our customers. Visit Aspen's Internet site for more information resources, directories, articles, and a searchable version of Aspen's full catalog, including the most recent publications: **http://www.aspenpub.com**
  Aspen Publishers, Inc. • The hallmark of quality in publishing
Member of the worldwide Wolters Kluwer group.

Editorial Resources: Jane Colilla
Library of Congress Catalog Card Number: 97-36248
ISBN: 0-8342-1006-1

Printed in the United States of America

1 2 3 4 5

*For my son Luke.
He spontaneously overcomes
restrictions of flawed architecture and design
every day.
May he always measure his abilities
as bounded by unlimited horizons.*

*This book is dedicated to the
Wizard's Wormhole Estate Family Collection.*

Table of Contents

Introduction vii

Chapter 1—Ergonomic Analysis Planning and Responsibilities 1
Defining Your Responsibilities 1
Identifying the Decision Makers in an Organization To Determine Ergonomic Analysis Format 2

Chapter 2—Assessing Needs Within an Organization 5
Criteria Selection and Baseline Statement 7
Assessing Ergonomic Need 9
Needs Assessment Criteria Entries 11
Subjective Comfort-Level Survey 11

Chapter 3—Gathering Background Information 13
Defining the Job 14
Physical Measurement 20
Workstation Measurement 21
Dynamic Force Measurement 23
Videotape Recording and Videotape Log 25
The Task Breakdown: Defining a Job in Detail 26
Task Cycles 31

Chapter 4—Risk Factor Identification 35
Recording and Describing Risk Factors 36
Risk Identification Procedures 50

Chapter 5—Intervention Discovery 53
The Importance of Creativity 54

Applying Creativity in the Intervention
  Discovery Process 57
The Intervention Worksheet 63

Chapter 6—Report Writing 65
Formatting Reports 65
Introductory Statement 66
Job Background and Description 67
Risk Factor Identification 68
Intervention Suggestions 69
Summary and Recommendation 70

Chapter 7—Planning an Intervention 73
Product Information and Comparison 74
Building an Intervention Resource File 74
Product Comparison Worksheet 75
Gathering Additional Information 75
Planning the Intervention 77

Chapter 8—Implementing Interventions and Reporting Progress and Effectiveness 83
Installation or Initiation 84
Orientation 84
Maintenance Considerations 85
Progress and Effectiveness Report 85

Appendixes 87
Appendix A—The Need for Appropriate
  Intervention Orientation 89
Appendix B—Intervention Considerations ... 91
Appendix C—Evaluating, Selecting, and
  Installing Ergonomic Interventions and
  Devices 99
Appendix D—Vendor Phone Numbers 103
Appendix E—Model for Comprehensive
  Ergonomic Program in an Office Facility 105
Appendix F—Suggestions for Creation of a
  Facility Ergonomic Coordinator 109
Appendix G—Worksheets 111

Index 209

Introduction

The effectiveness of an ergonomic analysis is measured by whether it results in a positive change in the work environment. The work that is involved in preparing a report—its length, the number of formulas used, digitized computer images, standards cited, research articles quoted—is irrelevant if the analysis does not motivate someone to create a positive change. The purpose of this book is to assist readers in using a "tool kit" approach and using their time to help facilitate that positive change.

The tool kit approach is intended to allow the user to focus his or her efforts on innovative problem solving by limiting time spent on analysis activities that will not enhance the chances of an intervention being implemented or the effectiveness of that implementation. By way of analogy, an individual may be highly skilled and capable of using all the tools in a woodworking tool kit. He or she will typically use only the tools that are necessary to complete a project. An effective ergonomist, although capable of detailed step-by-step analysis, will limit himself or herself to the tools necessary to complete the goal of creating a positive change in the work environment.

In most cases, the purpose for generating an ergonomic analysis and report is to provide someone with information. That information can be used to make certain decisions about either redesigning a workstation or making some change in how procedures are performed to create a safer and more effective work environment.

Sometimes, people think that there is one universally accepted format for providing that information. They think that a report must have a lengthy and scientific look. The unfortunate result is that creative and practical suggestions are buried in a cumbersome and ineffective document. An effective ergonomic analysis is one that facilitates, drives, and even compels individuals to act upon your recommendations. One of the keys to being able to create an effective ergonomic analysis lies in customizing the report to a format that most appropriately fits the organizational culture.

The most efficient use of your time will be facilitated by selecting and using the appropriate tools from this book. At the core of the tool kit approach to ergonomic analysis is the idea that you limit your reports to include only the essential information to make a decision regarding workplace modifications. This avoids unnecessary expenditure of your energy and effort and ensures the most efficient use of your time.

When used properly, the procedures in this book will result in the following:

1. effective communication of your findings
2. objective assessment of your suggestions
3. best chances for adoption of your suggestions for positive ergonomic modifications in the work environment
4. most efficient implementation of interventions for positive ergonomic changes
5. management and affected employees accepting and utilizing changes

WHAT IS THIS BOOK ALL ABOUT?

This book includes instructional guides, worksheets, report templates, model scripts, agendas, checklists, and menus. It is designed to provide all the necessary tools to perform ergonomic analyses. Master forms and models are provided for you. The worksheets provide options and simple methods for performing each step in an ergonomic analysis. You may refer to the book to determine the procedure for a specific step in the analysis process. You select the appropriate master form from the blank samples.

To perform the appropriate measurements and data gathering, some equipment may be required. The standard accessories are a stopwatch, vernier calipers, a folding ruler, a pen for marking still pictures, and a calculator. You may wish to augment these standard accessories with a force gauge dynamometer, a video camera, a stop-motion videotape player, and/or an instant camera. The intention of the analyst and requirements of the client should determine the specific components.

The process is simple and straightforward. First, select the appropriate steps. Next, select the appropriate tools or templates, or sets of procedures or forms necessary to complete each particular step. At this point, you may be wondering which steps are appropriate. The initial steps will help you identify the appropriate steps in the process. They involve assessing your personal commitment and then assessing the corporate culture or the willingness of the organization to act upon your suggestions. Each of the steps will become clearer as you learn about them and practice the skills presented in this book.

Think of this book as a workbook that is filled with various tools such as hammers, screwdrivers, saws, and pliers—only, instead of being those physical tools, they are actually various forms, templates, and sets of procedures. When you are getting ready to build something, you do not go to your toolbox, open it up, and look at every tool you have to try to determine how you could possibly build something using all those tools. You start out with a certain project in mind and then select from your toolbox only those tools that are necessary to complete the project. That is a simplified way of looking at a tool kit approach to turning out an ergonomic analysis. Use only those tools that are necessary for your ergonomic project.

WHAT ARE THE STEPS IN THE ANALYSIS PROCESS?

The analysis process starts when a facility determines that there is a need in a specific area. It does not end until that specific need has been addressed and evaluated. Although an analysis can be performed in a few days or weeks, the implementation of suggested interventions and the measurement of the effectiveness of those interventions may require several years.

Analysis Planning and Responsibilities

You assess your capabilities and limitations by defining your role in the ergonomic process over the next few years. In addition, the corporate culture of an organization is assessed. This assessment will include determining the decision-making process used by an organization. This will help you design the most effective way to perform an analysis and present the results of that analysis.

Assessing Ergonomic Need

To accurately measure the effectiveness of an ergonomic intervention, certain baseline data should be recorded. These data can also be used to assess the specific need for ergonomic analysis within a facility. The selection of the appropriate baseline data is determined by objectivity, availability, and practicality. It is also important to consider information gathered during the first step about the decision-making process and the organization's corporate culture and attitudes.

Gathering Background

The baseline data compiled can be used to select an area in need of ergonomic analysis. In this step, an accurate assessment of that work environment is developed. This involves defining the job in terms of duties and tasks, recording static workstation components and dynamic force demands, and often videotaping the performance of specific tasks.

Risk Factor Identification

A step-by-step comprehensive protocol is used to look at each body part and record specific types of risk factors. It is critical that nontechnical jargon be used to record and quantify each risk factor. This will facilitate better comprehension by all parties involved in acting upon this information.

Intervention Discovery

A creative critical thinking approach using a freeform model allows unlimited possibilities for workplace improvement regardless of your level of experience in this area. A comprehensive approach ensures that all aspects of the work environment are considered. This includes workstation design, workstation modification, changes in process design, and worker-based programs.

Report Writing

This is often the point when your information must be communicated to a decision maker in the organization. The clarity of this presentation can significantly affect the chances of your suggestions being acted upon. Appendixes A through F can be used to provide additional information to a client.

Intervention Planning

In this step, you consider the many requirements for implementing an intervention. This critical step in the ergonomic process is often missed. Lack of realistic plans for the implementation of an intervention has been responsible for failure of many otherwise valid and important interventions. An important aspect of this step is intervention research. In many cases, it is not necessary or possible to design a new workstation or device. Individuals with limited experience can search for the type of product in an intervention database and request information directly.

Implementing Interventions

It is important to detail the steps in implementation of an intervention to facilitate line management and employee acceptance and compliance. This includes verifying the delivery, set-up, orientation and training, and follow up.

Project Progress and Effectiveness

Most interventions require ongoing support to modify or augment the intervention as needed. This can be assisted by ongoing monitoring of the impact on ergonomic criteria selected in initial steps and by communicating with management.

Chapter 1

Ergonomic Analysis Planning and Responsibilities

DEFINING YOUR RESPONSIBILITIES

Before assuming the responsibility for performing an ergonomic analysis, you should first assess your capabilities and limitations by defining your role in the ergonomic process. The ergonomic process starts when a facility determines that there is a need in a specific area. It does not end until that specific need has been evaluated and addressed. Although an analysis can be performed in a few days, the implementation of suggested interventions and measurement of the effectiveness of those interventions will require several years. With that in mind, you are encouraged to start by determining which of the steps in an ergonomic process you will realistically be able to perform and which will require assistance from others.

In this first step, you examine how much time you have, what resources you have at your disposal, and how much effort you are going to be able to put into the process of creating an ergonomic analysis. As you use the various steps, you will learn how much time and effort are involved with each one as well as how to assess how important each step is in getting your suggestions accepted. This will make it easier to decide whether you will continue to perform those steps. Initially, this may be determined by what your experience has been, what your capabilities are, and your background in ergonomics.

Worksheet 1–1 (Responsibilities for an Effective Ergonomic Process) in Appendix G has been designed to assist you in the process of determining what roles you and other people will play in the performance of an ergonomic analysis. On this sheet, you will read many of the details that must be considered when performing an ergonomic analysis. Try reading through this sheet now to determine how many of these tasks you will be able to perform or how many of them might need to be performed by someone else. You may discover as you read these that certain details aren't clear at this point. Hopefully, as we go through the overall process, these details will become more apparent.

Exercise 1–1 Responsibilities for an Effective Ergonomic Process

Locate the worksheets in Appendix G. The responsibilities worksheet (Worksheet 1–1) will be included. Make a copy of the worksheet. You may want to modify your responsibilities occasionally. Follow the directions in that worksheet to help determine your role.

IDENTIFYING THE DECISION MAKERS IN AN ORGANIZATION TO DETERMINE ERGONOMIC ANALYSIS FORMAT

The purpose of this step is to identify the individual or individuals within an organization who are most likely to be able to make decisions regarding the implementation of interventions. Most organizations have at least one decision maker. Keep in mind that the decision maker is not always the individual who hired you or assigned you to perform an ergonomic analysis. If you are working as an outside consultant, you may be hired by a health and safety person or human resource person; however, that individual will simply take your report and pass it along to perhaps a facility manager who will then weigh the validity of your suggestions. The decision maker is the individual who can add items that may require more capital expenditures to a budget.

You can significantly enhance your chances of creating a positive change in the work environment by tailoring your analysis to suit the decision maker in an organization. If an individual would like a 100-page preliminary report followed by a 5-page summary, that format will most likely result in action. If an individual likes many illustrations, formulas, and citations from other organizations, then perhaps that's the type of information that should be included in your analysis. If you are fortunate, you will be dealing with an individual who will make his or her decisions based on a single page of suggestions. You will still have to perform several other steps in an ergonomic analysis process, but, ultimately, the only information that you'll be providing is that single page. The benefit of this approach is that, by tailoring your report to an individual or individuals, you're less likely to waste your time on unnecessary steps in an ergonomic analysis.

This step also will demonstrate the long-term aspect of an ergonomics project to everyone involved. This problem does not have a single easily recognizable solution. Purchasing new chairs for everyone does not result in an instant ergonomic work environment. The solutions or interventions will have to be assessed over the long term.

Try examining the specific corporate or organizational culture of the facility in which you're working. Try to determine the level of need for ergonomic intervention perceived by the decision makers in your organization. Try to determine how decisions have been made in the past and, most important, what was done to drive the decision makers to act. Although you may be able to speculate about this on your own, it is more efficient to meet with the decision maker. The information you get from that meeting will design the action plan for your analysis and act as a blueprint for your final presentation. Worksheet 1–2 (Ergonomic Analysis Format Question List) in Appendix G will help you to plan and structure that meeting.

It is also helpful to find out what the final outcome of an ergonomic intervention should be for it to be considered a success. Chances are, the decision makers in the organization had something in mind when you were given the project. They wanted to see something occur. It may have been a reduction in the incidence of injuries, the restoration of productivity outputs, or the issuance of a presidential award for employee concern. Your goal is to determine that final outcome.

This is a critical part of the process to ensure that you are going to effectively use your time by performing only those steps that are required for others to act on your suggestions for workplace improvement.

Exercise 1–2 Ergonomic Analysis Format Question List

For this exercise, you are to become the decision maker. If you are the manager of a facility or are self-employed, this should not be difficult. If you are a manager, imagine that you have been given the authority to make a decision regarding some type of ergonomics change in an area. Read the questions in Worksheet 1–2. Imagine how you would react to being asked each question. Rate the difficulty in answering each question from easy (1) to very difficult (5).

Keep this information readily available and refer to it when planning the order of the questions that the decision maker will be asked.

CHAPTER 2

Assessing Needs Within an Organization

In this step, you will determine the most appropriate data to be used, first as a criterion for measuring the need for ergonomic intervention and then as a baseline for gauging the effectiveness of modifications in that work environment. These data will be compiled, analyzed, and eventually reported.

To accurately measure the effectiveness of an ergonomic intervention, certain baseline data should be recorded. These data can also be used to assess the specific need for ergonomic analysis within a facility. The selection of the appropriate baseline data is determined by objectivity, availability, practicality, and the organization's corporate culture and attitudes regarding the need for an ergonomic process. In your meeting with the decision maker of an organization, you will certainly have formed an idea of what type of data or criteria he or she would like changed.

The science of ergonomics can be used to do the following:

- facilitate maximum productivity
- support consistent quality performance
- enhance worker comfort and long-term employee well-being

We can use this definition to help consider the different direct measurement options available for assessing need. You may choose to consider productivity variances. Productivity may be either higher than expected or lower than expected in a particular area. Quality variances can also be an indicator of ergonomic need. These can be assessed by checking reject or rework records (records of work needing to be repaired or replaced). Finally, injury incidence rate can help reflect the comfort and long-term well-being of employees. This can be determined or calculated by using an Occupational Safety and Health Administration (OSHA) 200 log or an in-house medical department visit log. Typically, three different individuals track this type of information, making the compilation of all this criteria somewhat awkward and, in many cases, impossible.

There are other types of measures of a need for ergonomic intervention that are more indirect. Turnover and absenteeism may point to a need for ergonomic intervention, but more often they are indicators of the relationship between the workers and management in an area. Of the two, turnover is probably the best indicator of need. A company may have an entry-level position that, in a way, tests new employees. This job may have the highest physical demands or be one of the more tedious jobs. Individuals will typically attempt to transfer out of this position as soon as they have the experience or seniority to do so.

Subjective comments made by individuals within an organization can be another way of determining ergonomic need. Often, everyone within a facility can point to one particular job that they all know to be the most difficult. The disadvantage of using subjective opinion is that measuring the effectiveness of interventions would be impossible. However, subjective information obtained through a standardized comfort-level survey could be used. The procedures for this step are straightforward and can be found in the subjective comfort-level survey discussed later in the chapter.

Three tools that can help you quickly assess and select the appropriate criteria for assessing ergonomic needs are Worksheet 2–1 (Criteria Selection and Baseline Statement, including how to compute incidence rates), Worksheet 2–2 (Ergonomic Need Criteria Comparison Chart), and Worksheet 2–3 (Subjective Comfort-Level Survey), in Appendix G.

Start by considering what criteria the decision maker in an organization would like to review to justify some type of change. You may wish to compile some additional information, so consider what other information is available, assuming that you have the time and cooperation of individuals who are, at this point, responsible for collecting these data.

CRITERIA SELECTION AND BASELINE STATEMENT

The Criteria Selection and Baseline Statement in Appendix G (Worksheet 2–1) can be helpful. This information will eventually be included in the introductory statement of your final ergonomic analysis. This form gives you an opportunity to document who is currently responsible for recording the information, who will be responsible for collecting it, how often they will collect it, and what the measurement is at this point.

Exercise 2–1 Calculating Incidence Rate

Ace Manufacturing has a total of 750 employees. 700 of them are full-time (40 hours/week). Fifty of them are part-time (20 hours/week). The plant closes for two weeks per year and everyone takes vacation at the same time.

Injury Record by Department

Department	Full-Time Employees, 40 Hours/Week	Part-Time Employees, 20 Hours/Week	Number of Injuries
JIT	10	0	4
DIST	35	15	6
GP	200	0	13
HR	5	0	1
TACK	30	5	2
SLT	125	0	4
LIP	135	25	15
ENG	15	0	1
EXT	55	0	2
LAM	40	0	5
QA	30	0	5
MAINT	20	5	3
Totals	700	50	61

1. Locate Worksheet 2–1.

2. Duplicate each of the forms in that worksheet.

3. Determine the number of hours worked in a year for each of the departments listed above, as well as for the facility as a whole.

4. Using the form showing how to compute incidence rate, determine the incidence rate for each of the departments listed above, as well as the incidence rate for the facility as a whole.

ASSESSING ERGONOMIC NEED

Another tool is the Ergonomic Need Criteria Comparison Chart (Worksheet 2–2) in Appendix G. Once you gather the information about the various departments in a facility, this worksheet allows you to compile it all in one chart. This is broken down into direct measurements such as number of injuries, productivity variances, rework and reject variances, and indirect measurements such as lost time days, turnover per 100 employees, absenteeism per 100 workers, comfort-level survey results, and number of cumulative trauma disorders. You can, of course, modify this chart to meet the specific culture of your organization. The worksheet allows you to expand or reduce actual numbers to reflect a department with 100 workers in it. It is unlikely that you will have this much data available on a regular basis.

Exercise 2–2 Needs Assessment

Several factors should be considered when assessing the need for ergonomic analysis within a facility. The following fictional injury records may be used when determining the priority order for jobs to be analyzed. Spend a few minutes looking through the injury records and then compare the different departments using Worksheet 2–2.

Select the three departments with the greatest need for ergonomic intervention. Be prepared to explain your choices to a decision maker in the organization.

Ergonomic Needs Assessment Comparison

Department	No. of Injuries	Productivity Variance	Rework/ Reject Variance	No. of Lost Days	Turnover per 100	Absent per 100	Comfort-Level Survey	No. of Cumulative Trauma Disorders
JIT	4	100%	100%	6	5	35	1.70	2
DIST	6	100%	100%	15	1	31	2.00	2
GP	13	105%	99%	71	3	15	1.90	6
HR	1	98%	100%	0	0	7	0.50	0
TACK	2	115%	97%	1	0.5	23	2.40	0
SLT	4	95%	85%	56	1	28	1.40	0
LIP	15	110%	90%	123	18	32	2.10	10
ENG	1	0%	0%	5	4	4	1.00	1
EXT	2	78%	85%	2	0	5	1.30	1
LAM	5	130%	100%	31	9	10	1.40	1
QA	5	100%	0%	9	0	7	1.10	2
MAINT	3	0%	0%	21	3	12	2.00	0
Totals	61	—	—	—	—			

NEEDS ASSESSMENT CRITERIA ENTRIES

- Number of injuries—Number of injuries in the department listed. Percent of injuries shows department injuries as a percentage of all facility injuries.
- Productivity variance—Percentage of target productivity goals; 100 percent means met target, less than 100 percent means produced less, and more than 100 percent means exceeded production goals.
- Rework or reject variance—A simplified measure of quality, reflecting how many products needed to be repaired or replaced; 100 percent means met target, less than 100 percent means poor performance, and more than 100 percent means better than expected performance (fewer rejects).
- Number of lost days—Total number of days lost as a result of an injury.
- Turnover per 100 workers—Number of people who transferred or resigned from that department. This number is adjusted to reflect a department with 100 workers in it.
- Absent per 100 workers—Number of days that people from the department were absent in the period being studied. This number is adjusted to reflect a department with 100 workers in it.
- Comfort-level survey—Empirical average level of comfort based on anonymous survey; 0 means no discomfort, 1 means discomfort rarely, and 2 means discomfort more often.
- Number of cumulative trauma disorder cases—The number of reported cumulative trauma disorder cases in the department during period studied.

SUBJECTIVE COMFORT-LEVEL SURVEY

The Subjective Comfort-Level Survey (Worksheet 2–3) in Appendix G provides a model for gathering and processing data. A subjective comfort-level survey helps to assign a numerical value to the level of discomfort at several body parts for individuals and groups of people. This can be helpful in supporting a decision to initiate an analysis in a particular area. In addition, it can be repeated every few months to get feedback on the effectiveness of workstation improvements once they have been implemented.

Exercise 2–3 Tallying Subjective Comfort-Level Survey Results

1. Locate and duplicate the comfort-level survey instructions in Worksheet 2–3.

2. Use the responses as listed below for a single department in a facility to calculate the comfort level for each of the body parts. Remember to separate responses of "mild" and "strong"—these responses should be evaluated separately. There is no way to combine the two in any meaningful way.

Low Back Mild Discomfort: 1, 1, 3, 2, 1, 1, 2, 2, 1, 4

Shoulder Mild Discomfort: 1, 2, 2, 2, 1, 4, 3, 1, 1, 1

Hand/Wrist Mild Discomfort: 2, 2, 2, 3, 1, 1, 1, 2, 2, 1

Low Back Strong Discomfort: 1, 2, 2, 2, 1, 1, 2, 2, 1, 2

Shoulder Strong Discomfort: 1, 2, 3, 2, 1, 4, 1, 1, 2, 1

Hand/Wrist Strong Discomfort: 2, 4, 4, 3, 1, 1, 1, 3, 2, 1

Chapter 3

Gathering Background Information

The purpose of gathering background information in the ergonomic analysis process is to describe and define the job that is being analyzed. This is done to assist in identifying and quantifying the physical demands of the job. This information can be used to compare the severity of various conditions or practices that may put the worker at risk. In addition, this baseline information will help you, later in the process, to determine the most effective job improvements. It will probably take a long time for workstation modifications to be implemented and an even longer time to measure their overall effectiveness. To accurately measure the long-term effects of job-related risk factors as well as the performance of worksite improvements, you will have to periodically confirm that the job demands have not changed significantly.

There are two parts of the gathering background information step. One involves defining a job and breaking that definition down into various job duties or tasks, and the other involves getting information on the physical surroundings of the workstation itself. The decision to perform some or all steps in this process will be determined by your available time, your experience, and whether you think the steps are essential to providing information to an organization's decision maker.

If you are working as an outside consultant, it may be necessary to complete this step in a detailed and comprehensive manner. By providing this information in the body of an analysis, you help to establish your credibility.

DEFINING THE JOB

Start by talking to someone who *knows* about the job. He or she can provide potentially useful information related to logistical consideration, such as transfers, meals, unions, training, and opportunities for advancement. You will have the chance to verify this information as you talk to several people within the organization. Next, talk to someone who has experience *doing* the job. Preferably this is someone who is currently doing this work, but a supervisor or someone who has done the job in the past can also provide much of the input for this section. When talking to people, it is best to use a conversational tone and not to sound like an investigator.

This book provides a tool for structuring this early contact that should assist you in organizing the way you gather information and record your observations as you perform your ergonomic analyses. Worksheet 3–1 (Introduction and Job Description) in Appendix G contains forms for recording background information, worker impressions, and job description and duty lists. These forms are designed to be used independently or in combination with each other. Using the worksheet can facilitate writing a narrative ergonomic analysis report or serve as a stand-alone record of your work.

Job Background Information

The job background information form in Worksheet 3–1 calls for you to identify the company's name, the company contact, the address, and the phone number. This is helpful if you are working as an outside consultant and will be mailing your completed analysis. The rest of the information on this form involves such things as the number of hours per day a person works, whether overtime is available and/or mandatory, whether a person works at the same workstation all day or moves about the plant, whether there is a production rate or quota that is officially required, whether the work is indoors or outdoors, type of equipment used, and training necessary. The human resources department or management may be able to provide some answers before you talk to any of the employees. You may choose to confirm the day-to-day perceptions and practices when speaking with employees.

Review the job background information form and then do Exercise 3–1.

Exercise 3–1 Job Background Information

Spend a few moments thinking about all the jobs you have ever had. Include summer and part-time jobs in your recollection. Think of the one that you considered to be the worst. Locate and duplicate the job background information form from Worksheet 3–1 (Introduction and Job Description). Fill out the form for that worst job.

Worker Impressions

The worker impressions form in Worksheet 3–1 gives you an opportunity to get input from an individual worker or workers. In general, you want to know how long they've been at that job, what they like best about it, what their perceptions of the duties are, and what they consider to be the toughest part. Often people will answer this last question with a response that relates to the management style of their boss or the newest cut in health care benefits. Although this information has little to do with ergonomics, it may help to further define the corporate or organizational culture. Ideally, you are attempting to get responses that address the physical demands of their job. In addition, the suggestions that employees have for changing their work environment are usually helpful.

Exercise 3–2 Worker Impressions

Locate and duplicate the worker impressions form in Worksheet 3–1 (Introduction and Job Description). Ask a coworker or friend about his or her job. Enter his or her responses on the worker impressions form. The practice of gathering this information from someone who knows and trusts you may help you prepare to talk to someone who does not know you.

Job Description and Duty List

The job description and duty list form in Worksheet 3–1 focuses only on the verbal description of the job and moves you through the process so that a list of duties can be created. Be certain to enter identification information such as the name of the company, where the analysis is being performed, the job title, the job number, who you are, and the date. Directly following that is an opportunity to write a job summary. This job summary should be a concise statement of the duties and responsibilities of the particular job being analyzed. If, in fact, you are looking at an entire area, then this would be the place to indicate that there are several different duties, jobs, and responsibilities existing in that area.

Review the example in Exhibit 3–1 and then perform Exercise 3–3.

Exhibit 3–1 Sample Job Description and Duty List

Company Name: Michael's Hospital

Job Title: Laundry Sorter   Job Number: 1656

Analyst Name: Burke   Date: 1/6/95

Job Summary

Laundry Sorter takes bags of laundry from the collection areas and places them in a transfer cart. Employee pushes cart to sorting room, empties bags, and tosses soiled laundry into the appropriate bin.

Exercise 3–3 Writing a Job Summary

Write a job summary for the job outlined below.

Ace Resale, Inc., has purchased a new product. They got a great bargain; however, they are not sure of the quality of the product. To make sure that the product is "up to spec," they have a department to inspect and re-pack each product before they ship it to the customer.

The inspector removes each small packaging box, disassembles the pieces, measures and weighs the oversize nut, and then reassembles the product and places it in the company's own box. In addition to this duty, he or she also is responsible for upkeep on all the personal computers in the plant. On average, he or she spends about one hour per day working on computers. For the last 45 minutes of the day, he or she will review all the paperwork of the other inspectors.

The inspector begins work at 6:00 AM. He or she has two coffee breaks—one at 7:30 AM and one at 1:30 PM—for 15 minutes each. He or she goes to lunch at 10 AM for one hour and goes home at 3 PM.

Job Exposure Calculation

Job exposure refers to the number of hours that a worker is directly responsible for working. This is determined by taking a normal shift and subtracting any coffee breaks, meal periods, or "down time" caused by equipment failure.

Review the example in Exhibit 3–2 and then perform Exercise 3–4.

Exhibit 3–2 Job Exposure Calculation

On-Duty Time: 10:59 PM
Number of minutes for first break: 10
Number of minutes for lunch: 45
Number of minutes for second break: 10
Any other down time (minutes): 15
Off-duty time: 7:59 AM

10:59 PM to 7:59 AM equals 9 hours
9 hours minus 10 minutes (first break)
    minus 45 minutes (lunch)
    minus 10 minutes (afternoon break)
    minus 15 minutes (other down time)
    equals 7 hours, 40 minutes job exposure

Exercise 3–4 Job Exposure Calculation

Use the job information in Exercise 3–3 to determine the job exposure.

Developing a Duty List

In this step, you will create a list of duties for this job title. A duty is a general statement describing an activity that is a responsibility of a worker. Examples include "loading trucks," "cleaning up area," or "performing paperwork." In addition to listing the duties, you are also given the opportunity to include the "duty exposure" or number of hours needed for each duty.

Sometimes it may be difficult to determine what a typical day is. In these cases, you should continue the analysis using a given day and one example of how much of each duty is performed. Once you have saved this information, you can change the job identification numbers and repeat the process to reflect a different day. It may be helpful to observe a series of "typical days" that will reflect the extreme of exposure to each duty. Modifications to previously entered information can be made at any time. It may be necessary to estimate the number of hours that a duty is performed each day.

Read the example in Exhibit 3–3 and then do Exercise 3–5 to develop a duty list for the inspector in Exercise 3–3.

Exhibit 3–3 Sample Duty List

The Garage Door Opener Assembler is responsible for building and shipping garage door openers. To do this, he or she must get the necessary garage door opener parts from stock. This usually takes about 45 minutes at the beginning of the shift. Next, he or she assembles the garage door openers and then packs them. He or she spends about 5½ hours building the openers and ½ hour packing them up. The last 20 minutes he or she applies the address labels and postage.

Duty List

Duty No.	Duty Description	Exposure
1	Retrieving materials from stock	45 minutes
2	Assembling garage door openers	5½ hours
3	Packing garage door openers	½ hour
4	Shipping garage door openers	20 minutes

> ### Exercise 3–5 Developing a Duty List
>
> Read through the job information in Exercise 3–3 again and create a list of duties and exposures for each duty. Determine the percent of the workday spent performing each duty.
>
> 1. Make a list of duties.
>
> 2. Determine the duty exposure and percentage. Calculate the percentage job exposure for all duties listed using the following formula:
>
> Duty percentage = (Duty exposure/Total job exposure) × 100

PHYSICAL MEASUREMENT

You can begin gathering information about the physical workstation and environment. Worksheets 3-2 through 3-4 in Appendix G will help you to accurately and comprehensively compile this data.

Worksheet 3-2 (Workstation Measurement) provides a format for recording physical measurements including heights, depths, and distances. Worksheet 3-3 (Dynamic Force Measurement) provides suggestions and recording format for measuring forces associated with the target job. Worksheet 3-4 (Videotape Recording Log) gives instruction and hints for recording videotape and developing a log. The order in which you choose to gather information can be arbitrary, or it can be a subjective response to what is going on. It may be dictated by site factors, such as whether anyone is at the workstation.

WORKSTATION MEASUREMENT

The purpose of this step is to objectively and carefully record data about the physical description of the work environment. Such things as the workstation height and depth, the distance to reach for various controls at the workstation, clearances underneath the workstation, and adjustable components of the workstation itself or any chairs that might be there are measured. This information is not gathered for comparison with a "safe" or "ideal" standard. There are no known standards for workstation heights that have been proven effective in the long-term prevention of cumulative trauma disorders. This information is recorded to establish a baseline for modification or for comparison with other workstations in the future. The Workstation Measurement worksheet (Worksheet 3–2) can be used to help record this information.

Use an instant camera to take a picture of the workstation. The price of these cameras is often as little as $20. If possible, try to get a side view and a front or back view. For the purpose of consistency in all your pictures, shoot the picture at the operator's eye level. This is the simplest method for recording three-dimensional information without complex drawings and calculations. Find a spot within your picture from which you will be able to measure all the distances clearly. This will be your reference point. Everything will be measured from this point and recorded on the worksheet. This is much less confusing than measuring from different points within the workstation.

Draw lines directly on your photos and out onto the worksheet with a permanent ink marker (Figure 3–1). If no instant camera is available, you can take the picture with a regular camera and go back later, or sketch the workstation. This can be extremely time-consuming, and unless you are talented at drawing to scale, it can make your data much more difficult to use. One of the

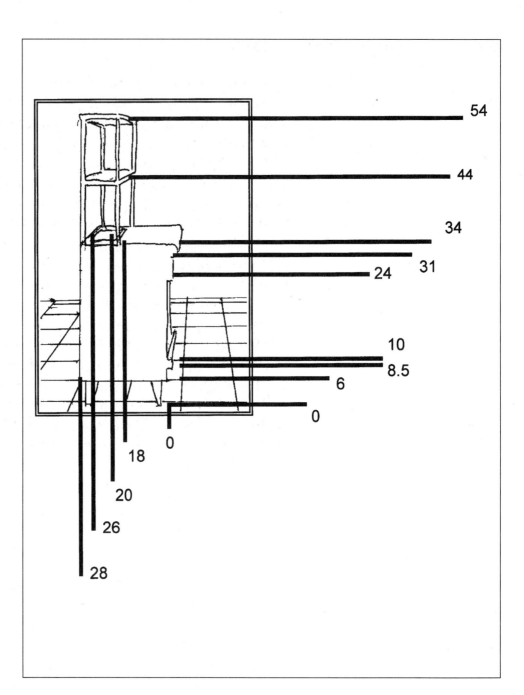

Figure 3–1 Workstation Measurement Record Showing Lines Drawn on Photograph

best times to perform the static workstation measurement is when no one is in the workstation. This can be before a shift, or after or during a meal break.

DYNAMIC FORCE MEASUREMENT

The next data-gathering process involves measuring the dynamic forces necessary in performing a particular task. Once again, these are not being measured to compare them with any set standards, because there are no set force standards. There are no known maximum lifting, pushing, or pulling forces that have consistently been associated with a reduction of cumulative trauma. To measure these, you may need a force gauge dynamometer, which is a device that measures the compression or tension or pulling force placed on it. These are available from several sources. For about $125 you can buy one that will measure up to approximately 60 pounds.

Worksheet 3–3 (Dynamic Force Measurement) will help in recording this information in an organized and comprehensive way. It will also provide hints for attaching your force gauge dynamometer to the object being tested. A template is provided on the worksheet for recording the information. Exhibit 3–4 shows an example.

Exhibit 3–4 Sample Detailed Measurement Worksheet

Workstation	Measured By	Date
Hospital hallway	Burke	4/22/86

Trial 1	Trial 2	Trial 3	Trial 4	Trial 5	Trial 6	Average	Units
35	37	44	36	35	35	36	lb

Method of attachment (if any): Bisected eye hook

Additional Comments

Force (no.)	Units (lb, kg)	Are Required To Push, Pull, Lift, or Hold (select one)	Task Description
33	lb	push	Pushing laundry carts to sorting area.

Draw diagram of measurement technique used or paste picture in space below.

VIDEOTAPE RECORDING AND VIDEOTAPE LOG

Videotaping the job is one of the most helpful practices that can be performed in an ergonomic analysis. If you are going to take the time and effort to videotape, it is advisable to take the time to write a videotape log.

Videotape will allow you to later closely examine and analyze not only what a person does but also whether any conditions or practices that potentially can harm him or her exist or are being performed. As a rule, it is best to start with an establishing shot as far away from the individual as possible. Try to show not only his or her whole body and the workstation but some of the surrounding area as well. After shooting some footage, turn off the camera and walk a bit closer until perhaps the camera has the individual's workstation and his or her entire body in view. Try to videotape an entire task cycle or an entire repetition of activities at this distance. Then get closer and perhaps videotape just the hands, the feet, or the upper body performing certain parts of the task. You may find it is helpful to perform this procedure from a different angle. In other words, you may first videotape from the right side, then from the left side, and then directly in front of the person. In actuality, the limitations in most work environments will dictate what type of videotape you will be able to shoot.

The Videotape Recording Log (Worksheet 3–4) provides a model for videotaping a target job and then setting up a written log. The log should be drawn up as soon as possible after the videotaping session. The form has places for entering the date, some way of identifying the information from the tape, who shot it, and the location.

Rewind the tape to the beginning using a tape player that you will have access to on a regular basis. Set the counter to zero and then start to watch the tape. Whenever anything significant changes, for instance the task being performed or the distance that you are from the individual working, stop the tape player, record the counter number, and then start the tape again. Do not reset the counter to zero. Continue to play the tape, entering any descriptive information in the description column, stopping the tape, and entering the counter number whenever possible. The more detailed you are, the easier you're going to find it later when you look for specific shots of certain tasks being performed.

Exercise 3–6 Videotape Recording Log Practice

Make a copy of the Videotape Recording Log in Worksheet 3–4. Use your personal videocassette recorder at home to tape a five-minute section of the nightly network news. Use a section with several news footage sections. Follow the directions on the worksheet to develop a videotape log for the five-minute section.

THE TASK BREAKDOWN: DEFINING A JOB IN DETAIL

Sometimes it is appropriate to describe a job in more concise terms than a summary statement or list of duties. There are several methods for doing this. Some systems call for defining a standard set of movements or activities. Each duty is then broken down into these predetermined actions. Other systems will reduce all activity to anatomic motions at various joints. These are typically extremely time-consuming and do not necessarily enhance the possibility of improving the workplace. A more practical approach is to break a duty down into tasks. The task breakdown approach allows the analyst to control the level of detail into which duties are broken down.

A task is a specific physical activity of a definable period during which a duty is performed. The only *requirement* is that it be a consistent and measurable unit of time. That means the task must have a beginning and an end and be repeated consistently or performed only once.

The following are examples of tasks:

- loading a pallet with boxes
- making an ice cream cone
- mounting a tire

One way to test the task you have defined is to identify the end product or unit of completion. If you cannot determine one clearly, then you may need to re-examine your breakdown. If a unit of completion is apparent, then you probably have performed the task breakdown accurately.

For the tasks listed, the units of completion might be the following:

- loaded pallets
- ice cream cones
- mounted tires

As long as the criteria you use for breaking each duty down into tasks are consistent and you state the unit of completion, your results will be accurate. There are several accurate and correct ways to describe the same activity. If the task description is "Placing boxes on a pallet," the units of completion could be loaded pallets or individual boxes.

The more detailed you are, the more descriptive the analysis will be. If you believe that a detailed analysis will be required to facilitate positive changes in the work environment, then proceed in that manner.

The first benefit of performing a task breakdown is that you will be able to state the number of times each task is performed each day. This number can also be called the quantity. For example, if Mary loads 50 pallets per day with boxes and the unit of completion is a loaded pallet, the quantity per day is 50.

This will provide a more detailed framework from which to continue the analysis, provide a basis for comparison among the various activities, and facilitate the quantification of ergonomic risk factors. This can also help meet Americans with Disabilities Act (ADA) job description requirements by providing a task list of activities to assist in the appropriate placement of job candidates.

There are other ways to describe and define each task. The simplest is to state that a task is performed for a certain number of hours each day. You have probably already done this for the duties when writing a duty list for the job. The number of hours can be referred to as the task exposure. For example, if Bill makes ice cream cones for five hours per day, the task exposure is five hours.

It may be helpful to describe the speed at which a task is performed. The speed can be called the rate. This rate is actually more of an average rate rather than an instantaneous rate. When you drive your car, the instantaneous rate will vary from moment to moment. Traffic or road construction may require you to slow down or speed up. Typically, you concern yourself with the average rate. When you arrive at your destination, you can determine your average speed as "miles per hour." When a task is performed a certain number of times for a specified period, you can determine the rate of completion. For example, if Mary loads 50 pallets per day for eight hours, her rate is 6.25 pallets per hour (50/8 = 6.25).

Exhibit 3–5 shows how to combine the descriptions to consistently describe a task.

Exhibit 3–5 Template for Describing a Task

<Task no. 1> <Task description> is performed
<Exposure> hours per day
at a rate of <Rate> <Units of completion> per hour
for a total quantity of <Quantity> <Units of completion> per day.

Review the example in Exhibit 3–6 and then try performing the exercises in Exercise 3–7.

Exhibit 3–6 Example of Task Description

> Mary Jones assembles intravenous tube sets. She places the edge of tubing into the plastic joiner. This task is performed for 7 hours and 15 minutes each day. The sets are spaced 1 foot apart on the conveyor belt that moves from left to right in front of her. Mary must assemble every other set, because there is someone else assembling the other sets. The belt moves an average of 9 feet per minute.
>
> What statements can be made about the duties Mary performs?
>
> Answer: Mary performs the duty of assembling intravenous tube sets 7.25 hours each day.
>
> What statements can be made about the tasks Mary performs?
>
> Answer: The duty of assembling tube sets consists of only one task. That task is assembling tube sets.
>
> What is the exposure for this task? (How much time is spent performing this task?)
>
> Answer: Mary performs the task of assembling intravenous tube sets for 7.25 hours per day.
>
> What is the rate for this task? (How fast is this task performed?)
>
> Answer: The sets are spaced 1 foot apart on the conveyor belt. The belt moves at an average of 9 feet per minute. Mary must assemble every other set. At 9 feet per minute, 9 × 60, or 540 tubes will pass in front of Mary every hour. She will handle only every other one, or 270 of them each hour. The rate is therefore 270 per hour. Mary performs the task of assembling intravenous tube sets at a rate of 270 per hour.
>
> What is the task quantity? (How many times does Mary perform this task each day?)
>
> Answer: The formula for calculating quantity is rate × exposure = quantity; therefore, 270 tubes per hour × 7.25 hours per day = 1,957 (rounded off). Mary performs the task of assembling intravenous tube sets 1,957 times per day.
>
> You can also combine all the answers into one statement.
>
> The task of assembling intravenous tube sets is performed 7.25 hours per day at a rate of 270 tubes per hour for a total quantity of 1,957 tubes per day.
>
> (Notice that the unit of completion, tubes, has been added to the previous statement.)

Exercise 3–7 Task Description

Henry Pitkin works in a high-volume used car lot. His job is preparing cars before they are sold. This involves washing, waxing, and applying a protective coating on each tire on cars that are traded in. This duty is performed for five hours each day. He prepares about six cars in that time. It takes about 15 minutes to wash each car, 25 minutes to wax each car, and 12 minutes to apply the coating to all four tires.

1. What are the tasks and the respective units of completion?

2. What is the quantity? (How many times does he perform each task?)

3. What is the exposure? (How much time does he spend performing each task? Fortunately, in this particular case, we have been given the actual performance time for each of the tasks. This allows us to determine the exposure time with greater accuracy.)

4. What is the rate? (How fast does he work?)

In the last exercise, you may have noticed something puzzling. If it takes 15 minutes for a person to wash one car, then couldn't that person wash four cars in an hour? (The instantaneous rate is four cars per hour.) Actually, he washes 6 cars in 5 hours, or 1.2 cars per hour. To avoid giving the wrong impression, it is helpful to add one more statement to the task description. The additional statement is the performance time. The task of washing cars that takes 15 minutes to complete is performed at a rate of 1.2 cars per hour for 5 hours for a total of 6 times per day.

The performance time does not always fit into the overall equation perfectly. The best way to determine the performance time is to time the performance of a single task. Repeat this timing process until you have determined an accurate and consistent result. If a person is performing a task quickly, you will be forced to count the number of times that a task is performed in a minute and then perform the arithmetic to determine a fairly accurate performance time.

In another example, a worker loads 50 pallets in one eight-hour day. This means that he or she loads 6.25 pallets per hour. You time this process and determine that it takes about 4.5 minutes to load one pallet. You see that 4.5 minutes times 6.25 pallets does not equal 60 minutes. Why does this discrepancy exist? Often, that is what the production manager wants to know. Various pauses, distractions, and adjustments add up to this amount of time and should be expected to occur.

Therefore, the comprehensive task description reads as follows. The task of loading pallets, which takes 4.5 minutes to complete, is performed 8 hours per day at a rate of 6.25 pallets per hour for a total of 50 pallets per day.

This breakdown process can be time-consuming. Be certain that all your effort is actually going toward helping to sell an intervention to the decision maker in your organization before spending too much time on it.

TASK CYCLES

Sometimes, as you try to break a job down into its specific tasks, you will discover that a repetitive cycle is involved. This repetitive cycle may involve several different steps, some of which are repeated several times from beginning to end of the task and others that perhaps are repeated only once. It can be helpful to break this task cycle down. The Job Task Breakdown worksheet (Worksheet 3–5) in Appendix G can assist in this process.

First determine what makes up a task cycle. The easiest way to do this is to start listing the tasks in the order that they are performed. Continue until the first task listed is repeated. Once that happens, you have defined the beginning and end of your task cycle. The Job Task Breakdown worksheet will help you to perform this often tedious task.

Enter the first task in the first box on the left. Enter the next task on the next line but start it one box to the right. Enter the next task on the next line and start it one box to the right. If a task repeats, enter it on its own line but start it in the same column as the first time you wrote it on the worksheet. This is a long process and is helpful if you cannot figure out the various repetitions in your head. You will quickly develop a shorthand method for entering tasks or sequences of tasks that repeat.

For example, the Ajax Baggie Company produces bags ("baggies") for nuclear waste. These bags must meet certain rigid specifications for them to be sold. The inspector retrieves the bags from the extrusion machine, inspects them, binds them together with a loose-leaf ring, and places them into a box. The boxes are then placed on a pallet. Exhibit 3–7 shows the task cycle breakdown for this example.

The numbers in the final row of Exhibit 3–7 are obtained by adding the number of times an entry starts in each column. They represent the number of times that each task in the task cycle is repeated.

Exhibit 3–7 Sample Job Task Breakdown Worksheet

Assembles box							
	Retrieves pile of 100 baggies from baggie machine						
		Inspects baggies					
			Places baggies on loose-leaf ring				
	Retrieves pile of 100 baggies from baggie machine						
		Inspects baggies					
			Places baggies on loose-leaf ring				
	Retrieves pile of 100 baggies from baggie machine						
		Inspects baggies					
			Places baggies on loose-leaf ring				
				Closes ring			
					Places in box		
	Retrieves pile of 100 baggies from baggie machine						
		Inspects baggies					
			Places baggies on loose-leaf ring				
	Retrieves pile of 100 baggies from baggie machine						
		Inspects baggies					
			Places baggies on loose-leaf ring				
	Retrieves pile of 100 baggies from baggie machine						
		Inspects baggies					
			Places baggies on loose-leaf ring				
				Closes ring			
					Places in box		
						Seals box and places in pallet	
1	6	6	6	2	2	1	

Chapter 4

Risk Factor Identification

This chapter will cover how to consistently and comprehensively identify ergonomic risk factors in a work environment. The term "ergonomic risk factor" has been accepted extensively, but it is important to realize that it is not synonymous with "cause of cumulative trauma disorder." An ergonomic risk factor is a condition or practice that can act as an obstacle to productivity, a challenge to consistent quality, or a threat to worker comfort, safety, and long-term well-being. Of those three interrelated areas of concern, identifying and modifying obstacles to comfort will have a positive effect on productivity and quality. Therefore, the risk factor identification process will try to document any obstacles to physical comfort, impediments to effective operation of a work process, and risk factors for the development of cumulative trauma.

For the purposes of adhering to conventionally accepted terminology, all of the three conditions listed will be referred to as "ergonomic risk factors." The process you use for identifying ergonomic risk factors should be as comprehensive and consistent as possible. It should consider all of the parts of the body and not simply focus on the most recently publicized cause of cumulative trauma disorder. It should use a procedure that can be performed by another individual and result in the same outcome or the same risk factors being identified. The procedure for identifying ergonomic risk factors considers all parts of the body and identifies various types of conditions, practices, or exposures. In addition, it uses a standard procedure and protocol for quantifying and documenting each one of the identified risk factors.

It will be up to you to determine how closely you adhere to the suggested format within this section. Although at first it may seem cumbersome, in the long run this procedure will save you an enormous amount of time in trying to construct individual statements to record the various risk factors observed. This system has been used for several years by hundreds of people with different levels of ergonomics experience and education. Thus far, no one has reported an ergonomic risk factor that could not be documented with this system.

RECORDING AND DESCRIBING RISK FACTORS

Most risk factors fall into the following categories:

- awkward range positions
- unsupported postures
- forceful exertions
- environmental conditions
- excessive physiological demand

Describing the Body Position

You do not need to study anatomy and memorize technical jargon. Any clear and consistent description will suffice. The following are some ways of describing every motion you are likely to observe.

Foot and Knee

- points toes/pushes up on toes
- pulls end of foot up

- turns foot in
- turns foot out
- bends knee
- straightens knee

Shoulders

- moves arm over head
- moves arm back and behind
- moves arm to side and over head
- moves arm across body
- slumps shoulder forward
- pulls shoulder backward
- turns arm in at shoulder
- turns arm out at shoulder

Hip

- pulls knee up toward body
- moves whole leg back at hip
- moves leg out to side
- moves leg across in front
- turns leg in at hip
- turns leg out at hip

Low Back

- bends at waist
- bends backward at waist
- twists at waist
- bends to side

Neck

- bends neck down or forward
- bends neck up or back
- moves head to side
- turns head

Elbow

- bends elbow
- bends elbow backward

Wrist

- bends wrist down
- bends wrist backward
- bends wrist (to pinky side)
- bends wrist (to thumb side)
- turns hand over (down or in)
- turns hand over (up or out)

Hands/Fingers

- closes hand into fist
- bends fingers backward
- spreads fingers apart
- bends finger down or in

Awkward Range Positions

An awkward range position occurs when a person moves a body part as far as it will go or close to it. This may sound vague and subjective, and is meant to be just that. There are no standards for what would be considered an awkward range position. All individuals have unique structures within their bodies that will determine what is an awkward range position for them. Some individuals can maintain a position comfortably that other individuals would find uncomfortable and potentially painful. In fact, you might discover that the awkward range position for your right arm would be different from that for your left arm.

Try holding a pencil with two hands behind your back; one hand goes up and over your shoulder and the other hand goes down behind your back below your armpit (Figure 4–1, left). Once you've done that, change positions so that the opposite arm will now be going over your shoulder, and the other arm is now bent down and back behind your back (Figure 4–1, right). You are likely to discover that it is much more comfortable doing it one way rather than the other.

Figure 4–1
Demonstrating an Awkward Range Position

When looking at an individual, ask yourself the question, "Might this person be uncomfortable if he or she assumed that position on a repetitive basis or for a certain length of time?" An example of an awkward range position is reaching the arm up above the shoulder level. Another example is having to bend the wrist all the way down or cock it up to perform a task such as using a computer keyboard resting on a table. An individual who is forced to twist at the waist to load boxes onto a pallet for shipment would be considered to be assuming an awkward range position.

The suggested procedure for recording an awkward range position is to first observe an individual performing a task. Start at the feet and see if you notice any risk factors at the feet, ankles, knees, hips, and so forth.

When you notice the worker moving a body part to an awkward range position, record the body part, the position, and any information available regarding the frequency of the awkward range position. This can ultimately be reported as a statement such as "Worker assumes an awkward range position at the (body part)."

You can expand this by describing the motion that the worker is assuming. For instance, "Worker assumes an awkward range position at the wrist—bends wrist up," or "Worker assumes an awkward range position at the low back—twists to the left."

A suggested recording format is "Worker assumes an awkward range position at the (body part) (position) X times per task."

Exercise 4–1 Awkward Range Position Identification

Refer to Figure 4–2. The subject is performing a lifting task that typically is performed once every 15 minutes for 6 hours each day.

1. Draw a circle around any awkward range positions you can identify.

2. Use the format suggested in the template to record your observations.

Figure 4–2
A Lifting Task

Quantifying Risk Factors

You can describe a risk factor more concisely by expanding the statement about how often or for how long the worker is exposed. The best small unit to describe is the task. If you have not already done so, determine the number of times that the task is performed. To quantify the exposure to this risk factor, multiply the number of times the risk factor occurs per task by the number of times that the task is performed each day (Exercise 4–2).

Exercise 4–2 Quantifying Risk Factors

1. Go back to Exercise 4–1 and determine the quantity for any of the awkward range positions identified.

2. Use the sample form below to record observations.

3. As you continue to perform risk identification exercises in this chapter, try to quantify any other risk factors you identify.

Job title/duty/task: Rotor inspection—lifting rotors onto inspection hood					
Type	Body Part	Position	Amount of Force	Environmental Conditions	Frequency/ Duration
Awkward range position					

Risk Factor Identification   41

Another common type of risk factor is an unsupported position. An unsupported position occurs when an individual holds a body part without moving it or resting it on anything for a period. The length of that period depends on the body part, the position it is in, and the individual.

When individuals hold an unsupported position, the muscles are forced to maintain that body part against gravity at a static length. This is considered particularly stressful. In addition, the joints of the body are exposed to a more concentrated compressive force in the connective tissue, all of which means that it is uncomfortable.

Assume a position similar to the one depicted in Figure 4–3. In this position, you will be maintaining an

Figure 4–3
An Unsupported Posture

unsupported posture at the shoulders, elbows, low back, and one knee. See how long you can maintain this position before starting to experience discomfort.

A more likely example is an individual who must stand and bend at the waist to wash dishes in a kitchen sink. Another example would be an individual who is forced to keep his or her arms suspended while keyboarding or playing a piano. Prolonged standing or prolonged sitting without a backrest is an unsupported position of the low back. The muscles of the low back must maintain the trunk in an upright position without moving.

People will rarely maintain an unsupported position if they do not have to. Usually, they will shift their weight, fidget, and wiggle or use another body part, such as the other arm, to support the unsupported body part. When an individual is so intent on the task he or she is performing, the individual tends to not pay attention to the way his or her body is feeling, and in such cases the individual will maintain an unsupported position. For instance, when someone is driving a car in the rain or a snowstorm, he or she will often lean forward and get his or her face a little closer to the windshield. (A foolish effort, but most people do it.) When the individual arrives at the destination, he or she gets out of the car and notices that his or her neck and back are stiff and sore. Most people assume this is due to "stress."

The suggested recording format for an unsupported position is similar to the way that an awkward range position is recorded, for example, "Worker maintains an unsupported position at the low back—bends forward while washing dishes." The following format is used: "Worker maintains an unsupported position at the (body part)—(position) X hours per day."

Exercise 4-3 Recording Unsupported Positions

Refer to Figure 4–4. The subject is performing an assembly operation that typically takes about five minutes to complete. This operation is performed 30 times per day.

1. Draw a circle around any unsupported positions you can identify.

2. Record your observations below.

Job title/duty/task: Assembling carburetor					
Type	Body Part	Position	Amount of Force	Environmental Conditions	Frequency/ Duration
Unsupported position					

Figure 4–4
An Assembly Operation

Forceful Exertions

A forceful exertion occurs when an individual moves a body part against resistance or maintains a body part in a static position against resistance. Examples of types of forceful exertions are lifting, pushing, pulling, and static holding. In most cases, that resistance will be either gravity or friction. If an individual is lifting something, the force that he or she has to overcome is gravity. If a person is attempting to insert or assemble two pieces of machinery, the resistance to overcome is friction. If an individual is pushing a cart, he or she will be attempting to overcome both gravity and friction.

Exercise 4–4 Identifying Forceful Exertions

Grasp a large book or brick in one hand, with the arm extended and pointing down. In this position, you will be applying a static forceful exertion with the hand and fingers. Slowly bend your elbow and lift the book up to your shoulder. In this motion, you continue to apply a static forceful exertion with the hand and fingers as well as a lifting motion with the elbow.

With your elbow straight, lift the book up in front of your body. In this motion, you continue to apply a static forceful exertion with the hand and fingers, as well as a static forceful exertion with the elbow, and you are also performing a lifting motion with the shoulder. Hold the book out straight in front of you and slowly bend forward slightly and then slowly rise. This action requires you to maintain a static forceful exertion at the hand, fingers, elbow, and shoulder. It is the low back that is providing the lifting forceful exertion.

Lastly, hold the book out straight in front of you and slowly bend and straighten your knees. Your legs are providing the lifting forceful exertion.

You may be wondering at what point a forceful exertion becomes a risk factor or what the limits are for a safe forceful exertion. There have been several attempts to develop a practical limit. There are so many variables in any forceful exertion that it is impossible to determine a universal limit. Thus far, no one has been able to consistently make the transition from scientific research to practical application.

If there are no set limitations or absolute safe standards, how can it be stated that this is a risk factor? The reason for including forceful exertion in an ergonomic analysis is that it will always offer an opportunity to improve the work environment. For instance, if a person must apply a pushing force of 50 pounds to move a cart and you are able to improve the design of the cart or initiate a maintenance program to decrease the amount of force necessary to push the cart, then you have facilitated a positive change in the work environment. Only by recording these various forceful exertions can you be certain to consider those possibilities.

In the example just given of the individual pushing the cart, most people would assume the greatest force is going to be associated with moving the cart. It is important to look for the nonmoving or static forceful exertion. In those cases, when an individual is applying a forceful exertion, he or she is at some point attaching himself or herself to whatever he or she is working on. In the example of pushing a cart, the static forceful exertion will occur at the hands. In another example, if a person is picking up a barbell, he or she has to use his or her hands to grasp that barbell—to attach himself or herself. If a person is screwing two pieces of work together, he or she is applying a static forceful exertion to hold on to the screwdriver with one hand and a static forceful exertion to hold on to the piece being worked on with the other hand. The forearm is moving, but if the piece is not held in place and the screwdriver is not held tightly, the pieces will never be attached.

When recording a forceful exertion, it is appropriate to identify the type of forceful exertion. The suggested recording format for a forceful exertion is "Worker applies a (type) forceful exertion to overcome (number of pounds) X times per task."

You will notice this differs from the previous two methods for recording risk factors.

Exercise 4–5 Recording Forceful Exertions

Refer to Figure 4–5. The subject is performing a deburring operation using a hand-held, trigger-activated disk grinder. This is performed for four hours per day.

1. Draw a circle around any forceful exertions you can identify.

2. Use the following form to record your observations.

The following are *types of forceful exertions*: lifting, static, pushing, pulling, and twisting.

Job title/duty/task: Deburring metal casts					
Type	Body Part	Position	Amount of Force	Environmental Conditions	Frequency/ Duration

Comments:

Figure 4–5
A Deburring
Operation

Environmental Conditions

The fourth type of risk factor considered is an environmental condition. The environmental conditions that are of significance in an ergonomic analysis are such things as heat, cold, vibration, and hard and sharp surfaces. An individual exposed to this type of environmental condition may not only suffer some discomfort as a result of the exposure but may also exhibit some type of adaptive behavior. For instance, if individuals working at a desk are exposed to the hard, sharp surface of the edge of that desk, they are likely to shrug their shoulders or not relax their arms. This leads to an unsupported position of the shoulders. An individual working in an office that is cold is less likely to relax in a chair and use the low back support of the seat back.

The suggested recording format for environmental conditions is "While performing a particular task, duty, or job, the worker is exposed to (type of environmental condition)." The type of environmental condition could be hot, cold, vibration, hard, or sharp surface. The following format is used: "The worker's (body part) is exposed to (environmental condition) X times per task, or X hours per day."

Exercise 4–6 Identifying and Recording Environmental Conditions

1. Look over the following list of possible ergonomic environmental risk factors. (Remember, many of these qualify by virtue of being obstacles to comfort for the worker.)

- heat or hot surfaces
- vibration
- hard surfaces
- cold or cold surfaces
- sharp surfaces
- other

2. Review the Exercises in this chapter and complete the sample form below.

Job title/duty/task: Rotor inspection—lifting rotors onto inspection hood					
Type	Body Part	Position	Amount of Force	Environmental Conditions	Frequency/ Duration
Environmental					
Comments:					

Job title/duty/task: Assembling carburetor					
Type	Body Part	Position	Amount of Force	Environmental Conditions	Frequency/ Duration
Environmental					
Comments:					

Job title/duty/task: Deburring metal casts					
Type	Body Part	Position	Amount of Force	Environmental Conditions	Frequency/ Duration
Environmental					
Comments:					

Excessive Physiological Demand

The last risk factor to be considered is excessive physiological demand. This is going to vary significantly from one individual to another based on the individual's overall physical condition. Some individuals who are in good physical condition are likely to be able to function at a higher level of energy expenditure than those who are not. Therefore, the best way to identify areas of excessive physiological demand is simply to identify physical indicators of excessive physiological demand. These physical indicators could be profuse sweating, rapid or heavy breathing, pushing for support, or holding onto a body part.

The suggested recording format is to state that an individual may be exposed to an excessive physiological demand because he or she is exhibiting certain signs of physical exertion and then list whatever those signs are: "Worker exhibits signs or behavior that may be an indication of excessive energy demand."

The following are signs of excessive demand:

- profuse sweating
- inability to speak normally due to rapid breathing
- reaching for support
- massaging or rubbing body part

RISK IDENTIFICATION PROCEDURES

The procedure for comprehensively identifying ergonomic risk factors is to focus on one particular type of risk factor, for example, awkward range position. Then, starting at a person's feet and working up through the knees, thighs, low back, shoulders, arms, and so forth, see if there are any awkward range positions that are being demonstrated. Then consider unsupported postures. Once again, consider the feet, knees, hips, low back, and so forth. When you've completed that, focus on looking for forceful exertions, again at the feet, ankles, and so on. Lastly, look for exposures to environmental conditions at the various parts of the body. Although this may seem somewhat cumbersome at first, you will discover that with practice you can work expeditiously and effectively using this procedure. It is suggested that at first you focus on individual risk factors while looking at the entire body, rather than focus on various risk factors while looking at the body one part at a time.

Once you have completed this body-part-by-body-part examination, then look for any signs of excessive physiologic demand. Exhibit 4–1 illustrates how this process can work. The Identification of Risk Factors worksheet (Worksheet 4–1) in Appendix G will help you to remain familiar with various types of risk factors and the suggested recording formats.

Exhibit 4–1 Risk Identification Procedures

	Assumes Awkward Range Position	Maintains Unsupported Position	Applies Forceful Exertion	Exposed to Environmental Conditions
Foot/Ankle				
Knee				
Hip				
Low Back				
Neck				
Shoulder				
Wrist/Hand				
				Signs of Excessive Energy Demand

Before getting started, you must decide whether you will be looking at each task and identifying risk factors or if you simply will be watching someone performing several tasks that make up a duty and identifying risk factors. In some cases, you may decide just to watch a person doing his or her job and try to identify as many risk factors as you can. Ideally, a task-by-task, body-part-by-body-part risk identification procedure will result in the most comprehensive list of risk factors. It will also result in an extensive list of risk factors.

Once you have completed your comprehensive list, you will probably go back and try to identify some key risk factors. Although several attempts have been made by various individuals to quantify which risk factors are the most serious, no one has thus far been able to develop any system that has held up or has been demonstrated by objective research. Therefore, when examining your list, look for those things that jump out at you. You will probably find you're identifying 20 to 30 different risk factors for each task or activity. Out of that, three or four will likely seem most excessive. In addition, watch for risk factors in various everyday activities that occur around you, whether at the supermarket or at a sporting event.

Chapter 5

Intervention Discovery

The next step in an ergonomic analysis is by far the most challenging and also the most enjoyable. You will attempt to come up with as many suggestions for positive ergonomic changes in a work environment. A positive ergonomic alternative is anything that will lead to increased comfort for the operator, increased safe and efficient operation, or reduction or elimination of risk factors for cumulative trauma.

This step in the process allows you to use your imagination, creativity, ingenuity, inventiveness, and any other creative talents you may have. In this part of the ergonomic process, individuals who thrive on breaking rules will do exceptionally well. Individuals who believe there is always an easier, better, and more effective way will get an opportunity to put their ideas into action. Individuals who think that there is a single correct answer to all solutions or believe that there is always going to be a correct and incorrect way to achieve a certain solution may find this to be the most difficult part of the process. Individuals whose minds tend to race, overflowing with new ideas when given some type of challenge, and who often have difficulty writing down all of the ideas or thoughts that they have will excel in this part of the process.

The goal is to come up with as many different ways to create a positive change in the work environment as you possibly can. It is important to realize that during this initial phase you should not be searching for a single right answer. This is not the time to even consider what is practical or what will fit into the specific corporate culture.

THE IMPORTANCE OF CREATIVITY

The significant problems we face cannot be solved at the same level of thinking we were at when we created them.

Albert Einstein

Meeting the challenges found in the work environment requires a different level of thinking—a new, fresh, and creative way of thinking. Being creative means looking at things differently, forgetting about the rules, and trying not to look for right answers.

The traditional standards and formula approach was used to design the work environment that is being analyzed. This approach demands that you look beyond a simple arithmetic calculation or limits found on a chart.

Remember that there are no instant answers. If a workstation modification is implemented, the long-term effect must be monitored for years. This is especially true if the criterion for success defined in the early steps of your analysis was a reduction in workers' compensation costs and incidence of injury. The ideas that a person creates and the ability to communicate those ideas are what are important. It is not necessary to know how to build an innovative device or workstation accommodation. Most businesses or organizations have the internal resources in their own engineering personnel or in their machine shops to take care of the construction requirements in a way that coincides with available funding.

Try Exercise 5–1, and ask other people to try it. One of the advantages to using creativity in the intervention discovery process is that it helps to get other people excited about a project. People want to be a part of a truly innovative idea. They want to talk about it and show it to other people. This is one way of increasing your chances of getting your creative ideas for modifying a workstation implemented.

Exercise 5–1 Correct the Equation

$$5 + 5 + 5 = 550$$

Use one line to make the equation above correct.

Answer: In many cases, the first answer that people come up with is to simply place a slash through the equals sign. This is the correct answer. Or perhaps it is only *a* correct answer. Try looking at the problem again to see if there are any other possibilities.

What if you drew a line connecting the top of one of the plus signs to the left arm of the plus sign? This transforms the plus sign into the number 4. This makes the statement correct in a way that is a little more interesting.

The first exercise showed how important it is to keep looking for answers even when it appears you have solved the problem. If you had stopped after placing a slash through the plus sign, you would have missed the much more creative solution.

Exercise 5–2 shows what happens when you look at something differently.

Exercise 5–2 Tic Tac Toe

Complete the game of tic tac toe to the right. You are "O" and it is your turn. What move must you make to ensure that you will win? You have about 30 seconds to complete this game.

Now you will make one change and see how it affects the way you think. What if the object of the game is to *not* get three in a row? (Normally, of course, the winner is the person who gets three in a row.) For this game, the first person who gets three in a row loses!

You are "O" and it is your turn. What move must you make to ensure that you will win, that is, that "X" will be forced into getting three in a row?

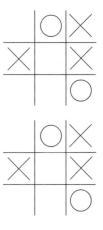

Now that you are in the mood, try applying your creative problem-solving skills to Exercise 5–3.

Exercise 5–3 The Boat and the Bridge

A narrow canal runs through the small town of Cropsey. A pedestrian bridge connects East and West Cropsey. One day, a large boat comes down the canal. The smokestack is too tall to pass under the bridge. The boat is owned by Silas Elliott, an eccentric multimillionaire. Assume that he is willing to do almost anything to get the boat to the other side. Your job is to come up with as many options as possible.

1. Start by brainstorming and think about anything that comes to mind. Examples are "raise the bridge" or "lower the river." Remember that Silas Elliott is eccentric, so do not limit yourself to things that are necessarily reasonable. Take about 45 seconds for this part of the exercise.

2. Focus only on the bridge. What can be done with the bridge to allow the boat to pass? Take about 30 seconds for this part of the exercise.

3. Now think about the boat. What can be done to get the boat to the other side? Take about 30 seconds for this part of the exercise.

4. Finally, consider the water. What can be done to the water to get the boat to the other side? Take about 30 seconds for this part of the exercise.

APPLYING CREATIVITY IN THE INTERVENTION DISCOVERY PROCESS

In the last exercise, you observed how your initial attempts to think of as many ideas as possible led to the creation of a list of suggestions. Then, by focusing on a specific aspect of the problem or the environment that you were dealing with, still with a creative and open mind, you were able to come up with more and more ideas. The same approach can be used in attempting to create positive ergonomic alternatives in a work environment.

The process for creative ergonomic problem-solving will require that you look at work environments from three different perspectives. The first category considers ideas that depend on the worker or worker participation. These are referred to as worker-based programs. The second category considers the process that makes up the specific task that is being performed. The third category, by far the largest, involves the physical environment or the workstation itself. Within each of these categories there are several subheadings to help you to focus your creativity and your problem solving.

Worker-based programs can be broken down into such things as training programs, exercise programs, rotation, personal protective equipment, procedural mandates, or even behavioral monitoring programs.

Process interventions consider the input and output of a process. The input includes raw materials, the position of those materials, and the packaging or level of completion of those materials. The output includes the finished product itself, the level of completion, and the physical placement of that output.

Workstation interventions consider the physical work environment and break it up into several questions regarding how the physical environment may be improved. For instance, ask the following questions: Can the workstation be adjusted or can it be made more adjustable? If something is being moved, can it be assisted to move? If something is being held in place, can it be assisted to be held in place? Can the number of times a task is being performed or the repetitions be decreased? Can the worker be supported?

Often, the best way to illustrate how this process works is to give an example of a task and then discuss how using this procedure can help you to come up with an extensive list of positive ergonomic alternatives. The sample task is unlikely to occur; however, the simplicity of it will assist in the demonstration of how to use this particular procedure.

For example, a fictitious safety product manufacturer and distributor offers a quarterly all-day seminar in various aspects of safety to its customers. Several speakers are invited. They are required to bring their own instructional handout materials. A worker is given the job of taking those handouts and punching holes in them so that they may be placed in a three-ring binder for distribution to the participants. The tool that this individual is given is a pliers-type, single-hole punch.

The objective of the exercise is to come up with as many positive ergonomic alternatives as you can. So, if we used the procedure outlined, the first step would be to consider worker-based programs.

The first of the worker-based programs would be a training program. What type of information should be provided and in what format could a training program be offered to this individual to help reduce exposure to the various ergonomic risk factors that are likely to be encountered? Examples of training programs would be a live, interactive training program where he or she would be taught the least stressful work practices for the operation of the hole punch. He or she might also be taught how to recognize the need for maintenance in the hole punch, or how to properly sharpen the hole punch. He or she might also be taught to take periodic breaks while performing this task.

Another alternative would be to have the worker perform exercises periodically while doing this task. It is important to recognize that the exercises this individual would perform should be specifically designed for the task being performed. In other words, an exercise program that would stress squeezing a tennis ball would probably serve to increase exposure and decrease overall comfort. An exercise program that would help the worker to perhaps resist opening his or her hand would probably result in the worker being much more comfortable. The frequency of performing these exercises as well as the amount of time spent per day performing these exercises can either be dictated or can be left up to the individual on an as-needed basis or when prompted by some type of timing device.

A rotation program may also be helpful for this individual—he or she would either share the performance of this particular task with other individuals or just limit the amount of time that the task is performed continuously.

Determining the least stressful work practices may also be useful, for instance determining that the wrist should be maintained in a specific position and then mandating that the individual adhere to this policy. This type of mandated procedure is effective only if individuals adhere to that procedure. This can often be difficult to enforce. One other way of possibly enforcing it is to establish some type of monitoring program. Examples would be to attach a wire to the individual's wrist that perhaps would beep whenever the wrist was moved into a stressful position, or simply, to monitor the number of times that individual moved his or her wrist into that position. The overall effectiveness of this type of program is questionable and may have some negative effects on the individual's attitude regarding work.

Another area to consider that involves the worker has to do with providing personal protective equipment.

Types of personal protective equipment related to ergonomics are various wrist splints, lift belts, and cushioning devices. It is important to recognize that a device that restricts motion may actually cause more problems than the original cumulative trauma it was attempting to prevent. Some individuals contend that a piece of personal protective equipment such as a lift belt will act as a reminder to facilitate adherence to a least stressful work practice. This is a new and unproved area of ergonomics.

The next group of interventions to be considered is the process types of interventions. Examine the input and consider if changes could be made and then examine the output and consider if changes could be made. The input in this particular process would be the physical papers themselves and the packaging the handouts come in. So the question comes to mind, "How might you modify the input to reduce the stress on the individual with this task?" It seems that the most apparent possible solution would be to ask the presenters to provide their information on three-hole paper. As an alternative, perhaps, the presenters could send in master copies that could simply be reproduced on three-hole paper. The papers with the three holes punched in them are the output of this particular process.

One way of reducing the stress on the individual who has to punch the holes would be to use an alternate binding method, perhaps a spiral binding method or simply placing the papers inside a folder. You might even consider asking the individual presenters to provide their information on computer disk and simply distribute the files to the attendees.

The last, greatest, and, for the most part, most fun category has to do with the changes in the workstation itself. Therefore, you ask the first question, "Can the workstation be made adjustable?" Well, if in fact this individual will be sitting, then it would be helpful to provide an adjustable chair. In some cases, even if a chair does not have adjustable features, you can still make it adjustable by adding cushions or bolsters. The same can be said for the workstation itself. The desk can be raised, either by placing blocks underneath the legs of the desk or table or by placing large, flat items on top of the desk or table. You might even look at the hole puncher and consider if it would be possible to adjust its size, or you may want to have hole punchers of different sizes available so the worker can select the one that feels the most comfortable. To carry this example further, you might consider having a hole puncher with an adjustable spring tension on it. The individual could then adjust the spring tension to be sufficient to open the jaws after it has been used to punch a hole, but not so great as to require unnecessary force to close the jaws.

The next consideration in the workstation category has to do with helping something to move. However, the only thing moving in this particular task is the hole puncher. Therefore, you might consider adding a small motor to the hole puncher to assist in opening and closing the jaws. As an alternative, perhaps, putting in a smaller spring to reduce the resistance of closing the jaws of the hole puncher would be helpful.

The next consideration is how something can be held in place. Well, the hole puncher is being held in place by the individual's hand and the papers are being held in place while they are being punched. Therefore, you might consider providing a clip to help hold the papers in place. A more practical suggestion would be to screw or attach the hole puncher to the workstation so that the individual would now feed things into the hole puncher and simply push down on the hole puncher.

The next consideration involves decreasing repetitions. One way this could be achieved would be to have the individual punch holes in as many pages as possible. This would result in reducing the number of times the jaws had to be closed. Another more obvious suggestion has to do with providing the individual with a three-hole puncher or attaching three single-hole punchers together with some type of bar.

The next workstation consideration is supporting the worker. Once again, you may want to consider providing the worker with a chair, or perhaps attaching armrests to the table to allow the worker to relax his or her arms while performing this particular task.

Another aspect of modifying the workstation involves consideration of environmental issues. In this case, the hard surface of the hole puncher can be softened by covering it with padding. Another way of helping to move the hole puncher would be to attach longer handles to it to allow the individual to use greater leverage.

Another environmental condition or environmental consideration would be to have the hole puncher made of a softer material or a less conductive material such as plastic rather than metal to keep it from getting too hot or too cold.

It is also helpful to consider the controls with which an operator will be interacting. Can a different system work better? For instance, you could replace the rotary channel selector on a television with an "up button" and a "down button" on a remote control. Sometimes this is as simple as making the existing controls larger or adding a ball at the end of a rod.

Another consideration for a workstation modification involves the physical access to the materials. If, in fact, the handouts are delivered in a box, you can make this easier by cutting the side of the box, opening the box, or emptying the box first. The other consideration involving access is that the papers must be fed into the hole puncher jaws. You can make this easier either by having the jaws open wider or by attaching a set of guides on the front of the jaws of the hole puncher to help place the paper between the jaws.

Another consideration has to do with the information input. In this particular case, that information is the placement of the hole puncher on the paper. The following examples, though somewhat ridiculous, illustrate this point: you might consider attaching a magnifying glass or a light to the end of the hole puncher or simply providing a light or magnifying glass on an articulating arm to help place the hole puncher on the paper.

THE INTERVENTION WORKSHEET

The Interventions To Create a Safe and Effective Work Environment worksheet (Worksheet 5–1) in Appendix G is designed to help you through the same creative thinking process just demonstrated. Try applying it to Exercise 5–4, which asks you to compile a comprehensive list of all the positive changes possible. Use a copy of Worksheet 5–1 to guide you through the exercise.

Worksheet 5–1 should be helpful in your initial attempts to compile a comprehensive list of positive alternatives. At the end of the worksheet, there is a list of possible options. This is not meant to be a comprehensive list of all the answers; it is meant only to help you facilitate your own creative thought processes.

Exercise 5–4 The Lawnmower

Curtis Hoofenheista has a new job. He mows lawns using a gas-powered lawnmower. To operate the lawnmower, Curtis must hold onto a kill switch bar on the handle of the lawnmower. It is not a power drive unit, and therefore he must push it to make it go.

Locate and duplicate Worksheet 5–1 to develop a list of positive ergonomic alternatives for Curtis. Take as much time as you need. This exercise may take you an hour or more to complete.

Now that you've seen how this process can be used to come up with an extensive list of interventions, you're likely to discover that one of the most challenging aspects of applying these procedures to actual workstations is continuing to have that childlike naiveté that facilitates the creative process. The more you know about a job and the more familiar you are with the organizational culture, the more difficult it is to suspend your belief that individuals would never adopt the ideas that you came up with. It is not unusual for people, in the midst of a creative, problem-solving frenzy, to suddenly be stopped dead in their tracks when presented with a workstation that they are familiar with. Therefore, work diligently to develop the ability to pretend that you know as little as possible about a particular process.

Once you have mastered this creative problem-solving skill as it relates to ergonomics, you will find that it is much more effective to compile an extensive list of positive ergonomic alternatives first and then review that list to determine which ones are most likely to be accepted or effective for the particular work environment that you are studying. You may also notice during this process that some of your more ridiculous ideas will give way to practical alternatives.

CHAPTER 6

Report Writing

Up to this point, you have gathered a lot of information. It seems appropriate to have a brief discussion on presenting this information in an effective format. This will vary greatly depending on whether the information is being presented by an outside consultant, an in-house committee, or someone else. The first step is to determine the purpose of the report. With this purpose in mind, you can customize the format and length to ensure its effectiveness. Regardless of the length of an ergonomic analysis, the report is likely to have approximately five essential components, as follows:

1. an introductory statement
2. a job background and description
3. a risk factor identification component
4. suggested interventions
5. a summary and recommendation section

FORMATTING REPORTS

Worksheet 6–1 (Templates for Reporting Ergonomic Analysis Results) in Appendix G offers several models for each of these components.

The introductory statement identifies the name of the job or the area that has been analyzed and briefly describes why the analysis was performed. The job background and description component of a report will identify the specific responsibilities, duties, or tasks associated with the job. The third section, the risk factor identification component, describes the ergonomic risk factors, conditions, or practices that potentially can reduce the productivity, performance, comfort, and safety of a particular workstation. The fourth section, the suggested interventions section, describes what interventions can be implemented to reduce the risk factors. The fifth section consists of a summary and recommendations. It describes what the greatest need is for interventions in that area and how modifications can be achieved. A more detailed discussion of each of the sections follows.

INTRODUCTORY STATEMENT

In the introductory statement, you should at least identify the name of the job or area that has been analyzed and demonstrate your understanding of the reason for that analysis. Additional information that may be provided in this area would be the name of the company, the name of the contact, and any information regarding the criteria used to select this particular job, or baseline information that has been gathered for this particular job. That baseline information may have been derived from an employee comfort-level survey or it may be productivity variance, rework or reject variances, or incidence of injuries.

Additional information offered will be the number of shifts, number of people working in an area, and perhaps any other ergonomic analyses or ergonomic interventions that have been implemented in that area. The following example shows how brief this section can be. This section may range in length from one or two lines to several pages. Your initial contact with the organization for whom this analysis has been performed will dictate the length of this section.

JOB BACKGROUND AND DESCRIPTION

This section basically answers the question, "What does the subject of the ergonomic analysis do?" The purpose of defining this is to establish a common understanding of the job function. This can also vary in length from just a couple of lines generally describing the responsibilities and duties of the individual to a comprehensive multiple-page breakdown of the task. The information for this part of the report is likely to come from the job background worksheets filled out during that part of the analysis. Depending on the needs of the facility, this section would probably contain the following elements:

- Job exposure—How much time does the subject spend working each day?
- Duty list—What are the duties that make up this job?
- Duty exposure—How much time is spent performing each duty?
- Task breakdown—What are the tasks that make up each duty? At what rate is each task performed? How much time is spent performing each task? How many times is each task performed each day?
- Narrative statement—The narrative statement is the brief job description containing information from Worksheet 1–2 and Worksheet 3–1.

The following example will demonstrate how this information may fit together.

On February 9, 1990, an ergonomic analysis was performed at the Lake View Nursing Home. The Lake View Nursing Home has been providing intermediate and long-term nursing care for approximately 120 individuals for the past 35 years. Information in this report was provided by Henry Pitkin and Dr. Howard Rowdie. Lately, there has been increased incidence of work-related injuries, primarily among the younger, less experienced workers. These injuries have been located primarily in the shoulder and neck. Dr. Rowdie has commissioned this report to identify conditions or practices that may lead to these types of injuries. He would also like any suggestions for modifying the workload for these individuals as well as any additional suggestions for preventing these injuries. In the past, the facility's physical therapist has offered yearly low-back injury-prevention courses. No objective data were maintained to measure the effectiveness of that training. Because the highest incidence of injuries occurred in the nurse's aide group, this group was targeted for this analysis. There are approximately 32 individuals employed as nurse's aides at this facility. An around-the-clock, three-shift operation is used. Nurse's aides are responsible for assisting nurses in daily resident care. This may include assisting in transfers, feeding, dressing, and walking. The exposure to these various duties will vary significantly from one day to the next and from one shift to the next.

During your analysis you probably would have gathered significantly more information from the employees at this particular facility. It is important to remember that you do not need to write down everything. The purpose of the section is to establish an understanding of the job and job functions that have been analyzed. Additional information may be kept on file for future reference or to support any suggestions that may be challenged.

RISK FACTOR IDENTIFICATION

The third section of the report will cover the identification of various risk factors. The purpose is to indicate conditions or practices occurring at the workstation that can act as obstacles to maximum productivity and performance or worker comfort and long-term well-being.

This book has offered specific definitions for each of the risk factors. Regardless of whether you use this format or a different one, be certain to explain any risk factors that you point out. It is important to be as consistent and comprehensive as possible. For each risk factor identified, you should state at what part of the body it is occurring and how many times a day an individual is exposed to it. In addition, explain any technical language or anatomical references cited.

This book offers a standard format for identifying and reporting each one of these risk factors. Depending on the needs of the decision maker of an organization and your time, you will have to determine if you are going to report all of the risk factors occurring for each task and for each duty or as an overview of the many tasks that a person performs.

Regardless of the approach taken, it is important to state what occurs when that individual is exposed to each risk factor. For instance, if you state that a person assumes an awkward range position at the wrist it is important to document what that individual is doing at the time he or she is exposed to that risk factor. The statement would be something like, "While stirring the slew depository, the crank stamper assumes an awkward range position with a wrist—the wrist is bent back. This occurs approximately 500 times per day."

INTERVENTION SUGGESTIONS

The fourth part of an ergonomic analysis report will cover the interventions that you have identified. The purpose here is to provide a comprehensive list of interventions. This should represent all the work you did. List a range of possible interventions, from something as simple as distributing a leaflet with least stressful work practices to redesigning a facility or even slowing down a line.

This book can be used for compiling this type of report. Once again, you consider the unique organizational culture of the facility to determine how detailed this section should be. You have the option of identifying the various interventions by simply referring to a particular job title, by specifying that an intervention would assist in the performance of a specific duty, by stating that an intervention would assist in the performance of a specific task, or even by specifying an individual risk factor that would be addressed by the initiation of a particular intervention.

Appendixes A through F can be used as models for providing additional information to a client. Appendix A discusses the need for appropriate intervention orientation. Appendix B describes factors to consider when purchasing particular kinds of equipment. Appendix C discusses the evaluation, selection, and installation of ergonomic devices and interventions. Appendix D presents a list of vendors' telephone numbers. Appendix E describes a comprehensive ergonomic program in an office. Appendix F suggests that the client create the post of ergonomic coordinator for the facility.

If the intervention plan described in the next chapter has been completed, it may be inserted in this part of the ergonomic analysis report. Because such research is usually time-consuming, it usually takes place after the initial report has been reviewed.

SUMMARY AND RECOMMENDATION

The final section of the ergonomic analysis report is probably the most important. In the summary and recommendation section, you will attempt to point out the most serious risk factors and the things that you think should be done as soon as possible. In many cases, this may be the only section a decision maker reads.

Therefore, it is important to be as concise and descriptive as possible to persuade the decision maker to act upon your suggestions.

You may find it helpful to go back through your analysis and highlight specific statements that you think are the most important. You can then use those statements to construct this final section.

In most cases, this final summary and recommendation should be no longer than one page. If you think more explanation is necessary, you can always refer to sections that have preceded this final summary and recommendation section. Remember, the format and tone of this section must be suited to the specific corporate climate and the requests of the decision maker.

Also, although this may be difficult, it is important to outline the anticipated impact that your suggestions would have on whatever criteria were used to either establish the need for ergonomic analysis or as a baseline to measure the ongoing effectiveness of an intervention.

Chapter 7

Planning an Intervention

Thinking through the many requirements for implementing an intervention is an important step in the ergonomic process. Many otherwise valid and important interventions have failed because of lack of realistic plans.

This process also helps in deciding which interventions are most likely to work within your particular organizational structure. In the previous intervention discovery step, you enjoyed the freedom to be as imaginative as possible. This step allows you to assess the practicality of those ideas.

By carefully planning and documenting the steps involved in implementing an intervention, you will find it is much easier to convince the decision maker in an organization that the intervention is possible. You will do this by indicating that you have considered the practical realities associated with the suggested intervention. You will also clearly define what action you want taken by the decision maker. This also gives you an opportunity to communicate the need for additional resources to implement an intervention, such as assistance from other individuals within the organization. The decision maker often requests this type of planning procedure at an initial meeting.

This process consists of two parts. The first involves gathering descriptive product information and then comparing the various features of the products. The second part considers the logistics of acquiring, installing, and implementing the interventions.

PRODUCT INFORMATION AND COMPARISON

Worksheet 7–1 (Building an Ergonomic Intervention Source File) and Worksheet 7–2 (Intervention Comparison) in Appendix G will help you to gather and compare information on various ergonomic interventions. This will help you to determine whether an intervention needs to be designed or if the same result can be achieved by using an existing product or service. During the research process, you are likely to discover several other interventions that may also help you to solve some of the problems in your facility. Once you have gathered this information, you will be able to compare the various costs and features from the many products and suppliers. It can be helpful if you are building your own personal ergonomic intervention resource file. Once you have contacted the various vendors by mail, you may want to contact them again to get more specific information regarding their products or services.

BUILDING AN INTERVENTION RESOURCE FILE

The question is "How do I know what type of ergonomic interventions are available and where can they be purchased?" To help answer this, a special section of this book deals exclusively with intervention research. Worksheet 7–1 helps you to find sources, and gather and categorize descriptive product and service information. A letter template is included for requesting information. You only have to add your name to the information request letter and send it to vendors you are interested in. A form for recording these requests and the information received is also provided.

PRODUCT COMPARISON WORKSHEET

The next step in the process is to compare the various products or services on which you have gathered information. Be sure to use only the information that is available in print. You may expect some inconsistencies in the way the products are represented (for example, data using certain units or standards may be included in one piece of product information and an entirely different set of data may be used in another piece of product information).

You may then want to discuss pricing, delivery dates, and other individuals who have used those products with vendors. Worksheet 7–2 is a model for you to use to compare various features. You may want to add columns of your own or make additional notes. The worksheet includes an intervention vendor questionnaire that may be helpful in compiling a list of questions to ask vendors.

Be candid. Disclose whether or not you have buying power and recommendation power. Often, individuals may request the name of the person who will be the decision maker in this process. For the most part, it is never a good idea to give out that name.

GATHERING ADDITIONAL INFORMATION

There are several methods for getting additional information about the various interventions as well as what would be involved in implementing each one. One of the best ways is to contact other organizations or facilities that have implemented this type of intervention. If you're part of a large corporation, you can contact other facilities within that corporation. You might also consider contacting facilities outside your own corporation. There tends to be a spirit of cooperation between individuals who have health and safety responsibilities, even among the fiercest rivals.

You will often find it helpful to talk to several different vendors. Each vendor will offer one or two ways to view the particular product or program he or she is selling. Additional vendors will offer various methods of evaluating, planning, and implementing the interventions. After speaking with several vendors or suppliers, you will have many different ways to consider a product and the various steps involved in planning and implementing it.

One of the most effective ways to get information from vendors is to ask them to compare their products to others that you may already be familiar with. Suppliers will often give you the names of other facilities or organizations that have used their product. Take advantage of this to contact those other organizations and get information about their experiences.

You may also want to consider asking ergonomic consultants for their opinion or assistance. Most experienced consultants will view this as an opportunity to demonstrate their expertise. This could lead to future business for them. They may not give you all the answers, but they will be in a position to provide some input. A consultant who is reluctant to give you some of his or her impressions or who insists on visiting your facility first is probably not familiar enough with your particular industry or situation to give you valid information. Confident ergonomic consultants, however, will welcome your inquiry and will probably supply you with some helpful information.

You may consider contacting one of the universities currently offering ergonomics programs. You may be able to contact advanced students eager to share what they know or to be involved with a "real" consultation.

You might also consider contacting the Human Factors in Ergonomics Society.

The federal government can be helpful in this area. Under the Freedom of Information Act you can get copies of local or national Occupational Safety and Health Administration (OSHA) citations. This will allow you to read the specific recommendations that OSHA made after an ergonomic citation. If you have access to library facilities, you may consider asking for a literature search using key words such as cumulative trauma, ergonomics, and the specific industry in which you are involved. Hopefully, this literature search will produce various case histories that would provide you with a structure from which to plan your intervention implementation.

Finally, you might consider researching how other programs or products were introduced within your organization. These do not need to be ergonomic or even safety types of programs or products. Consider the decision-making criteria used to select that particular product, how the product was implemented, how the product was accepted, and the final outcome of that implementation.

PLANNING THE INTERVENTION

Worksheet 7–3 (Intervention Implementation Plan) in Appendix G helps you to prepare a detailed plan for the implementation of an intervention. A detailed plan may not be necessary for the initial report; you may consider sketching out the planning process only for those interventions that you think will work within your particular organizational culture. Expect that many changes will occur within your plan. Make sure you communicate that the plan is a suggestion and can be modified. If you are not familiar with the operation of an organization (that is, if you are working as an outside consultant), it is best to be general and offer several options as part of the plan.

The following steps are suggested for planning an intervention:

1. presenting an introduction
2. researching and acquiring the intervention
3. comparing the research results
4. ordering the intervention
5. initiating installation and orientation
6. determining cost
7. measuring and reporting effectiveness
8. requesting action

The intervention implementation plan worksheet lists several detailed questions pertaining to each of these steps. Answers to the questions in that worksheet were used to produce the following example (Exhibit 7–1).

Exhibit 7–1 Sample Intervention Implementation Plan

INTRODUCTION

The overall productivity and performance as well as the comfort and well-being of the individuals in the order-taking department can be significantly enhanced by providing these individuals with new adjustable seating. There are approximately 40 workers in this area, and several have complained of ongoing low-back discomfort. There has also been a concurrent 10 percent reduction in productivity in this area. Based on our analysis, we believe that the provision of adjustable seating will result in a much more comfortable environment, reduce the incidence of injury, and restore productivity to its anticipated level.

RESEARCHING AND ACQUIRING THE INTERVENTION

Bill Jones and Mary Smith of the human resources and facility planning departments, respectively, have agreed to be responsible for researching various chair vendors. Several vendors have indicated that they are willing to have their chairs tested on site before a purchase commitment.

Several individuals will be recruited to be a part of this test. Each of them will undergo approximately 20 minutes of orientation regarding the operation of the seating as well as certain adjustment guidelines. After a three-day testing period, they will submit a written summary of their impressions.

Mary Smith will also be contacting other companies that have purchased these chairs in the past to get their impressions. Mary will ask questions regarding the overall effectiveness of this seating, specifically its impact on safety, incidence of injury, and productivity at other facilities.

COMPARING THE RESEARCH RESULTS

There are four reputable chair vendors available locally. Mary Smith will be responsible for arranging for the testing and analysis of the chairs. Bill Jones will be responsible for getting bids on the number of chairs that we potentially would be purchasing. Thus far, no bids have been submitted. However, other individuals have indicated that the cost per chair can be anywhere from $250 to $450 per chair.

ORDERING THE INTERVENTION

Each vendor has indicated that all the chairs could be delivered within a 6-week period with the standard 60-day payment terms.

continues

Exhibit 7–1 continued

INSTALLATION AND ORIENTATION

It is anticipated that approximately 30 to 40 minutes will be necessary to orient each of the workers regarding use of the new chairs. Before the delivery of the new chairs, Mary Smith will spend some time orienting the managers in the order-taking department about the new chairs. She not only will cover the information that will be provided to the employees but also will discuss the arrangements for any maintenance needs that could occur after the delivery of these chairs.

Either Mary Smith or some of the employees initially used in the testing process will be responsible for orienting any newly hired or transferred employees about the use of these chairs. This same group will also provide training for current employees when the chairs arrive.

COST

The cost per chair is expected to be between $250 and $450. The additional cost will be in the form of training time to orient workers regarding the appropriate and effective use of these chairs.

MEASURING AND REPORTING EFFECTIVENESS

Mary Smith will have the overall responsibility for tracking the effectiveness of these interventions. She already is responsible for maintaining the log of incidence of injury and lost time. The department heads will provide her with two pieces of information. The first piece of information will relate to the month-to-month productivity variances in that department. In addition, before the implementation of this particular intervention, each of the employees in that area will fill out an initial comfort-level survey. This will be used as a baseline to measure the overall effectiveness of these chairs as it relates to worker comfort. This information will be reported to the senior management staff on a quarterly basis.

REQUESTING ACTION

With permission, we would like to initiate the preliminary steps in the testing and comparison of various chairs available. An immediate response would be appreciated, because enthusiasm for this project is currently very high. Please feel free to contact me should you require any additional information about this suggested intervention.

A great deal of information was provided in Exhibit 7–1. In the cases when the information was not available, it was anticipated that it would be available after some initial steps in the research and testing procedure. Once again, your understanding of the unique culture in which you are working will indicate how much of the previous information is necessary in your organization.

Chapter 8

Implementing Interventions and Reporting Progress and Effectiveness

Once upon a time, the owner of a family restaurant called Rockbottom's presented his cooks with a brand-new "ergonomic" can opener. The owner, R.C. Rockbottom, had determined that using the new can opener was necessary and important and would affect him personally. He thought it was necessary because he had observed the cooks struggling with the old one. He thought it was important because he was concerned that the cooks might hurt themselves if they continued to use the old one. And lastly, providing the new can opener made him feel good about himself. It may have been because he thought that he was showing concern for his workers. It may have been that he thought it was a good business move to increase productivity. Or maybe he just liked to buy new kitchen gadgets.

For the new can opener to be effective, however, it has to be used by the cooks. The cooks need to see that it is necessary and important, and that it affects them personally. To facilitate this recognition, a *smooth* implementation of this new procedure is required.

A smooth implementation helps employees and line supervisors accept and comply with an intervention. The smooth implementation of an intervention can be summed up in the following three steps:

1. Verify all logistics associated with the installation or performance of the intervention.
2. Plan the employee and management orientation.
3. Plan for the maintenance need of the intervention.

INSTALLATION OR INITIATION

The installation should be performed by a qualified technician. It is a good idea to have at least one line worker present who is uniquely and completely familiar with the year-round operation in the facility. He or she may recognize conditions that may not have been considered and ask questions that may have been overlooked. In addition, the line worker will have the opportunity to learn many maintenance and set-up procedures and operational hints.

Be certain to have the installing technician actually operate all the functions of the intervention before he or she leaves. Next, have someone else from your facility operate all the functions of the intervention while the installer is there.

ORIENTATION

Plan the orientation in advance. Worksheet 8–1 (Implementation Orientation Agenda and Checklist) in Appendix G will assist in planning the orientation in a way that will help everyone see the positive advantages of the new intervention. Do not depend on the vendor to provide adequate training. If the vendor offers training, determine whether it complies with the requirements outlined in Worksheet 8–1.

You can reduce stress associated with any change in operation if you can make arrangements for immediate questions and concerns to be addressed in the first few days or weeks. Designate a person to field those questions immediately.

MAINTENANCE CONSIDERATIONS

Even though you have a new device, it is important to consider the future impact of this process change. If operational problems develop during the first six months of a new device's life, workers may revert to older and potentially stressful work practices. Therefore, it is important to meet the maintenance needs of the intervention in a timely manner. Arrange for any special training by in-house maintenance personnel. Establish a method for employees to communicate requests for maintenance. And, of course, set up a standard preventive maintenance schedule. The checklist in Worksheet 8–1 can be helpful in ensuring a smooth implementation for any intervention.

PROGRESS AND EFFECTIVENESS REPORT

The objective of this last step is to determine and report the overall effectiveness of an ergonomic intervention or project over time. This short report to management should accurately reflect ongoing ergonomic progress. It should, as concisely as possible, answer the question, "Is it working?"

Worksheet 8–2 (Tracking Progress and Effectiveness Report) in Appendix G is a valuable tool in quickly recording the effectiveness of an intervention. It provides a single worksheet for recording the changes to the specific criteria used to select and measure the impact of the ergonomic process. It can be used to help structure a report by providing a timeline. This can be helpful, especially in situations when the implementation of an intervention may have occurred months or even years after the initial analysis and recommendations occurred.

This report summarizing this information is likely to be most effective if presented in a narrative form. In general, it should be as short and concise as possible. It should reflect the efforts of all people involved in the ergonomic process and assign credit to key individuals.

Start by restating the fundamental information in a narrative format. That fundamental information would include the potential hazard identified at the workstation, the actions taken to correct the hazardous condition, and, finally, an answer to the question "Did the intervention work?"

Be as specific as possible regarding the outcome. If at all possible, put the results into a dollar amount. Include secondary benefits in a way that is as objectively measurable as possible. This might be stated as a reduced percentage in tardiness or absenteeism, specific unit per hour increase in productivity, or a reduced number of reworks per 100,000 units.

Additional information may be appropriate depending on the unique corporate culture. This might answer the following questions:

- What criteria were used to determine the workstations to be analyzed?
- When was the analysis performed?
- What were the possible interventions discussed?
- Why was this intervention chosen?
- Were there any complications in the installation or implementation of the intervention?

Appendixes

Appendix A—The Need for Appropriate Intervention Orientation 89

Appendix B—Intervention Considerations 91

Appendix C—Evaluating, Selecting, and Installing Ergonomic Interventions and Devices 99

Appendix D—Vendor Phone Numbers 103

Appendix E—Model for Comprehensive Ergonomic Program in an Office Facility 105

Appendix F—Suggestions for Creation of a Facility Ergonomic Coordinator 109

Appendix G—Worksheets 111

APPENDIX A

The Need for Appropriate Intervention Orientation

It is critical to the success of any intervention that the workers accept and use the changes implemented. Management's decision to initiate any type of ergonomic intervention should be combined with a commitment of support. Failure to establish this commitment as part of the decision may result in a waste of personnel and financial resources. The following list suggests a course of action to help ensure the effectiveness of an ergonomic intervention:

1. Allow the users (the people who ultimately will use the intervention) to have input.
2. Ensure that the need for this intervention is universally understood.
3. Ensure that the users believe that this intervention will fill the need.
4. Provide adequate training regarding operation of the interventions.
5. Provide adequate guidelines for setting up the interventions.
6. Provide a short-term, immediate feedback process to answer questions regarding operations and guidelines.
7. Perform ongoing follow up to ensure appropriate use of the intervention.
8. Assess the effectiveness and communicate with all users.

Appendix B

Intervention Considerations

CHAIRS

Sitting in a chair takes a combination of low-back and trunk-muscle control combined with solid balance skills. If done properly, sitting in a chair can provide rest and support for the sitter and allow the performance of other tasks for extended periods. In an office environment, the chair can be the major determinant of worker effectiveness and comfort. The chair should fit the sitter's body and meet the requirements of his or her job.

Chairs are being offered with just about every conceivable adjustment capability. In this case, more may not be better. Ease of adjustment and the proper training are equally as important as the number of features in a chair. There is nothing so puzzling as someone sitting uncomfortably in a $450 ergonomic chair that has not been adjusted since the day it left the factory. The following guidelines are used to select chairs

A chair should allow adjustment of features while a person is sitting on it. When a person is able to adjust the chair from a seated position, he or she can instantly assess and reassess the comfort level. If a person must get up, walk around the chair, and, in some cases, hold one part of the chair while spinning another, it is much more difficult to determine the most comfortable adjustment. Also, because of the additional steps, the operator is less likely to periodically adjust the chair during the day or even over the course of a year.

A chair should allow at least three adjustments: seat height, backrest height, and fore and aft backrest position. Some people believe that the purpose of an adjustable seat height is to allow persons with different leg lengths to sit with their feet on the floor. Although this may have been true when the majority of paperwork was performed in handwriting, the presence and increasing usage of a computer have changed the importance of adjustable height. A properly adjusted seat height will allow the person to maintain the least stressful position of his or her neck, shoulders, arms, and lower back.

The backrest height adjustment feature is important to allow the operator to rest the muscles of the lower back by providing a structural support for the upper body and trunk—in other words, by making it as comfortable to lean against as possible. Many people will sit forward and not use the backrest for the majority of the day. It may be that their backrest is not adjusted to the correct height. Therefore, it does not offer a better "feel" and does not encourage the operator to use it.

The fore and aft backrest position adjustment compensates for the different length of workers' thighs. Most chairs today are designed with a comfortable waterfall edge meant to support the thigh without putting pressure under the thigh or pinching the sensitive underside of the knee (popliteal area). If a chair lacks this, the seat pan (length) may be too short. When this is the case, a large part of the thigh will hang over the edge of the chair. The seat pan length can, in some cases, be too long. When this is the case the back of the knee will be in contact with the seat pan edge.

Some additional features are also desirable. A forward seat tilt tilts the pan of the seat forward. It allows the operator to reduce the angle of the legs and lower back. In some cases, this will allow the operator to also shift the position of his or her pelvis into a more comfortable position throughout the day. More important, this feature allows the operator to shift some of his or her weight onto the legs and reduce the muscle activity of the back muscles.

A seat backrest tilt is important to allow variability in the position that a person maintains throughout the day. Someone may feel most comfortable with a straight 90-degree position of the seat backrest while keyboarding. He or she may prefer to have it tilted slightly forward when performing handwritten paperwork and tilted back when talking on the telephone. When a chair does not have this feature, the operator will usually assume these positions by moving away from the backrest. This creates an unsupported position of the lower back, which can lead to fatigue and unnecessary stress in sitting.

A seat backrest locking and free movement option will accommodate the individual who performs several tasks simultaneously and thinks he or she does not have the time to constantly adjust the backrest tilt angle. By having the backrest float with the worker, he or she can still have the same back support whenever he or she needs it.

GLARE SCREENS

The purpose of a glare screen (Figure B–1) is to reduce the reflected light bouncing off the screen and enhance the display from the video monitor. Considerations are the type of material the glare screen is made of, the way that it is attached to the monitor, and how easy it is to maintain.

Employee evaluation of various screens is critical to ensure their use after you provide them. Just about any vendor of office supplies or computer equipment will

Figure B–1
Glare Screens

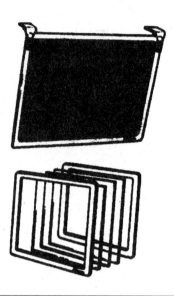

carry and sell video display terminal (VDT) glare screens or filters.

MONITOR STANDS

The purpose of a monitor stand is to position the computer monitor in a way that facilitates maximum comfort in the neck, shoulder, and eyes.

The variables to consider when selecting the appropriate monitor stand are mounting method, adjustable height, adjustable vertical tilt, adjustable horizontal rotation, and adjustable distance (eye to screen).

Most monitors can use several different mounting methods, including the following:

1. Sitting on the desk
2. Clamped to a work surface overhang on the workstation
3. Bolted into the work surface
4. Using a long clamp that hooks onto the top of the work surface and the bottom of the desk
5. Sitting on the floor

The purpose of tilting a monitor is to make it easier for the operator to see and to help control glare.

The purpose of adjustable monitor height is to allow the operator to maintain a normal, least stressful posture of the back, neck, and shoulders. If a librarian must bend the neck or move it forward or twist to the side to see the monitor, the muscles will be working all day. This constant working will result in pain and may contribute to the development of a cumulative trauma disorder.

Ease of operation and adjustment is important if several different librarians use the workstation or the workstation is used on different shifts. Most people do not readjust the monitor height often. Once it is set, it stays that way. If a computer is shared, then it will be important for the operator to be able to quickly readjust the workstation at the beginning of each shift. If desk space is not a consideration, then a stationary monitor stand that raises the level of the monitor and rests on the desk may be an option.

If desk space is an issue, it makes sense to remove the monitor from the desk and free up the space. Some type of suspension system will be necessary. For the purpose of definition, let's state that there are two types of suspension systems.

The post and arm system (Figure B–2) rides a post and has an arm attached to it.

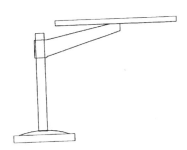

Figure B–2
Post and Arm

This will work in most workstations. The drawback will be attaching this system to the desktop. It will need to be screwed into the desktop because there is no space for a clamp to fit between the work surface and the cube wall.

Another type of monitor stand is the articulating arm. The articulating arm attaches to the desk or wall in a single hinge joint.

If the post and arm suggested previously are impossible to mount to the desk, you may be forced to consider a monitor stand that hangs from the top of the cube wall. This may create too great a distance from the operator and, therefore, should be tested. Also, this type requires a flat wall surface and, therefore, cannot be used in the corner of the cube, which is a popular location of many computers. The most likely solution will be the desktop monitor stand mentioned at the beginning of this section.

STANDING WORKSTATIONS

It is often advantageous to allow individuals the option of standing while working. This can be facilitated by having a workstation that can be raised high enough. Data tables come in many configurations and are typically raised either mechanically or electrically. You can also place the monitor and keyboard on a stand that can be raised to a standing work height.

DESKTOP EDGE PADDING

The purpose of desktop edge padding is to allow the operator to rest his or her arm more comfortably on the desk. The Ergo-Edge attaches to the edge of the desk and protrudes outward.

DESKTOP EASELS—DOCUMENT HOLDER

The purpose of a desktop easel is to angle the work being performed in a position that is most comfortable for the operator. Depending on the task being performed, it can be determined if a ledge on the bottom is needed to keep stacks of paper from sliding off the easel. You may also want to consider if the angle is infinitely adjustable or if you are limited to one or two angles. A device such as this can be easily fabricated by a worker in the machine shop within a facility.

ARMRESTS

When an individual is performing a task for a prolonged period on a static desktop, he or she may try to rest his or her arms completely on the desk top if there is room. An alternative is to use the armrest on the chair or even attach the armrest to the desktop. Several models have existed for a while. The newer models are considerably less expensive. Features to assess are ease of adjustment, independent height adjustment, tension adjustability, locking in place, and method for attaching to the workstation.

Appendix C

Evaluating, Selecting, and Installing Ergonomic Interventions and Devices

RESEARCHING DEVICES

List the minimum requirements for the device you want. This should be based on the duties the user performs, availability and delivery, price range, and input by knowledgeable ergonomic specialists. Consider the features that are most suited to the type of activity the users perform. Someone who uses one type of device for an entire shift will have different needs from someone who has other tasks to perform. This will help to decrease the number of product descriptions that have to be examined.

Contact vendors. A list of vendors can be obtained from the telephone directory, any ergonomics product buying guide, or an ergonomic consultant. The initial contact can be made either by phone or by mail. Tell the vendor the requirements for the device that you want, any price restrictions, any delivery date requirements, and the condition that you would like the opportunity to test the device on site.

Let the vendor know that you will be conducting a structured test and evaluation process and how long it will take. Some vendors do not want to leave a device with a potential client. Others will leave it with you indefinitely regardless of whether you buy from them. In-house "device evaluators" will need simple instructions.

You do not want the salesperson to be able to contact the evaluators because this could affect their decision. If each person who will use the device needs individual instruction, then the device is probably too difficult to use.

TESTING AND EVALUATING THE DEVICES

Use the information regarding features, availability, and pricing to select three or four devices that you think will meet the unique needs of the users. Narrow the choice to no more than three or four different devices.

Ask for volunteers to participate in the study. Each person should agree to use each device for a least two consecutive days. They will then be responsible for swapping with the appropriate person in the testing rotation. Each volunteer should be provided with ergonomic guidelines for setting up a workstation. In addition, each volunteer should undergo adequate training in the operation of each device. He or she should be given a specific form (Exhibit C–1) with questions for evaluating the device. This form should be completed for each of the devices. (Allow for subjective comments to be added regarding each of the devices.) Based upon the input of the testers, select the device best suited.

INSTALLING THE DEVICE

It is critical to the success of any ergonomic intervention that workers accept and use the changes implemented. You can significantly enhance the chances of worker acceptance and utilization by following a few guidelines such as the following:

1. Provide an orientation session for all the managers in an area.
2. Provide orientation to the workers who will be getting the new devices to ensure that everyone un-

Exhibit C-1 Sample Evaluation Form

Rate each of the following statements from 1 to 5. (5 means you agree completely, 1 means you disagree completely.)	Disagree				Agree
	1	2	3	4	5
I really liked the way the device felt when I . . .					
I had to adjust the device several times to . . .					
The device felt wobbly and insecure.					
I felt less tired at the end of the day.					
I think the device made me feel better at . . .					
I think the device made me feel worse at . . .					
I found the device easy to operate and . . .					
If I had this device, I wouldn't want anyone else . . .					
It fit well into, under, and around my . . .					
I liked the appearance of the device.					
This device is much better than the one I . . .					

On a scale of 1 to 10 (10 being the highest), I rate this device as _____.
List things that you liked about the device, as well as any concerns you may have or shortcomings of the device, on the back of this form.

derstands why the old devices are being replaced and why the new devices are better, knows how to operate the controls on the device, is familiar with ergonomic guidelines for setting up a workstation and how the new device fits into those guidelines, and knows whom to contact for additional information immediately as well as in the future.

3. Provide a mechanism for the short-term immediate feedback process to answer questions regarding operations and guidelines.

4. Post the instructions for device operation and ergonomic guidelines for workstation set-up.
5. Follow up periodically to determine the acceptance and utilization of the devices.

APPENDIX D

Vendor Phone Numbers

Adaptability	800-243-9232
Alimed	800-225-2610
Backcare Corporation	312-258-0888
BodyCare Inc.	800-858-9888
Boise Cascade	708-773-5100
Business and Institutional Furniture Catalog	800-558-8662
Cherokee Products Incorporated	800-535-9005
CompUSA	800-266-7872
Corel Corporate Seating	419-522-0001
Direct Safety Company	800-528-7405
Ergonomic Technologies	708-945-0009
Ergosource	800-969-4374
Global Computer Supplies	800-227-1246
Krames Communications	800-333-3032
Lyben Computer Systems	313-268-8100
MicroComputer Accessories	800-521-8270
Misco	800-876-4726
Moore	800-323-6230
National Business Furniture	800-558-1010
North Coast Medical	800-821-9319
Office Depot	800-685-8800
Office Furniture Center	800-343-4222
Office Max	800-752-0726
Personal Health Designs	518-458-8998
Quill Office Products	708-634-4800
Steelcase	312-321-3551

APPENDIX E

Model for Comprehensive Ergonomic Program in an Office Facility

This report contains several possible suggestions for improving the effectiveness and safety of many of the workstations. Each suggested intervention, if implemented individually, will decrease ergonomic risk factors and help to create a safer, more effective work environment. However, a more comprehensive and systematic approach will significantly increase the overall effectiveness and performance of the workers in those areas.

The following steps constitute a model for a systematic and comprehensive ergonomic program.

1. Determine policy regarding short, periodic exercises or rest breaks.
2. Determine what, if any, retrofitting equipment suggested in this ergonomic analysis will be offered to hourly workers.
3. Determine if hourly workers will be allowed to use retrofitting equipment that they produce at home or purchase independently.
4. Determine the role hourly employees will have in the evaluation, testing, and selection of any equipment or furniture obtained in the future.
5. Determine if assistance for workstation modification such as moving computers or adjusting desktop heights will be available.

6. Determine and document procedures for requesting and receiving workstation modification assistance.
7. Decide to provide a training program to employees that will do the following:
    - Outline what equipment will be available.
    - Describe procedures for acquiring any equipment.
    - Offer policies regarding use of equipment purchased or produced independently.
    - State policy regarding the performance of exercises.
8. Initiate training program:
    - Review contents of training.
    - Schedule supervisors' orientation.
    - Determine method for training; for example, voluntary (during business hours, nonbusiness hours, or lunch) or mandatory (by physical area, by department, or as determined by department supervisor).
9. Provide supervisor orientation:
    - Outline overall or systematic process.
    - Demonstrate content of employee training.
    - Define role of supervisor in the process.
10. Schedule hourly worker training.
11. Initiate subjective comfort-level survey.
12. Provide training to hourly workers.
13. Initiate new equipment or furniture selection process.
14. Determine roles and responsibilities of all persons to be involved.
15. Determine method to be used for testing, evaluation, and selection of equipment.
16. List sources for equipment.
17. Request information from sources.
18. Select vendors and equipment.
19. Arrange for on-site testing either on a loan or single purchase basis.

20. Provide workers or other individuals to be involved in on-site testing and evaluation of trial equipment with an outline and protocol for evaluating each product.
21. Compile results and determine products to be purchased.
22. Take delivery on all products.
23. Determine inspection procedure for each end-user.
24. Provide inspection procedure and check list to ensure that each end-user can operate all features.
25. Provide any additional orientation or training to each user.
26. Provide an immediate mechanism for initial questions or concerns regarding the new equipment or furniture.
27. Monitor effectiveness using quarterly subjective comfort-level survey and Occupational Safety and Health Administration (OSHA) and company injury incidence information.
28. Continue awareness through periodic review sessions, handouts, or posters.

APPENDIX F

Suggestions for Creation of a Facility Ergonomic Coordinator

Ergonomic coordinators will provide the following in-house support:

- Ongoing training and orientation
- Keeping up with new developments in the field
- One-on-one workstation adjustment upon request

This will require recruitment and designation of specific individuals at each facility. These individuals must be comfortable with training and dealing with people on a one-on-one basis. The selected ergonomic coordinators would require a short, intense training session that would include the following:

1. Background on cumulative trauma disorders
2. How to identify obstacles to comfort
3. General information on training as behavioral modification
4. Procedures for performing an employee ergonomics program

Ergonomics coordinators should be provided with several tools to perform their functions. These tools should include the following:

1. Video instruction on workstation setup
2. Video instruction in exercise performance
3. Video examples of employee training session
4. Creative techniques for program customization
5. Forms for office workstation analysis
6. Identification of risk factors
7. Possible interventions
8. Intervention sources
9. Step-by-step procedures for program design at their facility to do the following:
   - Determine facility policies
   - Set up in-house procedures for training, equipment modifications, and retrofitting equipment requests
   - Provide training for management
   - Provide training for employees

APPENDIX G

Worksheets

Worksheet 1–1 Responsibilities for an Effective Ergonomic Process: Instructions 113

Worksheet 1–2 Ergonomic Analysis Format Question List: Instructions and Procedures 117

Worksheet 2–1 Criteria Selection and Baseline Statement: Instructions and Procedures 121

Worksheet 2–2 Ergonomic Need Criteria Comparison Chart: Instructions and Procedures 125

Worksheet 2–3 Subjective Comfort-Level Survey: Instructions and Procedures 129

Worksheet 3–1 Introduction and Job Description: Instructions and Procedures 131

Worksheet 3–2 Workstation Measurement: Instructions and Procedures 137

Worksheet 3–3 Dynamic Force Measurement: Instructions and Procedures 141

Worksheet 3–4 Videotape Recording Log: Instructions and Procedures 145

Worksheet 3–5 Job Task Breakdown: Instructions and Procedures 149

Worksheet 4–1 Identification of Risk Factors: Instructions and Procedures 153

Worksheet 5–1 Interventions To Create a Safe and Effective Work Environment: Instructions and Procedures 159

Worksheet 6–1 Templates for Reporting Ergonomic Analysis Results: Instructions and Procedures 173

Worksheet 7–1 Building an Ergonomic Intervention Source File: Instructions and Procedures 185

Worksheet 7–2 Intervention Comparison: Instructions and Procedures 189

Worksheet 7–3 Intervention Implementation Plan: Instructions and Procedures 193

Worksheet 8–1 Implementation Orientation Agenda and Checklist: Instructions and Procedures 201

Worksheet 8–2 Tracking Progress and Effectiveness Report: Instructions and Procedures 205

Worksheet 1-1 Responsibilities for an Effective Ergonomic Process: Instructions

Before assuming the responsibility for performing an ergonomic analysis, you should first assess your capabilities and limitations by defining your role in the ergonomic process. Examine how much time you have, what resources you have at your disposal, and how much effort you're going to be able to put into the process of turning out an ergonomic analysis. The following steps will help to assign responsibilities.

ANALYSIS PLANNING AND RESPONSIBILITIES

You assess your capabilities and limitations by defining your role in the ergonomic process over the next few years. In addition, the corporate culture of an organization is assessed. This assessment will include determining an organization's decision-making process and will help you design the most effective way to perform an analysis and present the results of that analysis.

ASSESSING ERGONOMIC NEED

To accurately measure the effectiveness of an ergonomic intervention, certain baseline data should be recorded. These data can also be used to assess the specific need for ergonomic analysis within a facility. The selection of the appropriate baseline data is determined by objectivity, availability, and practicality. It is also important to consider information about the decision-making process and the organization's corporate culture and attitudes, which is gathered during the first step.

GATHERING BACKGROUND

The baseline data compiled in the previous step can be used to select an area in need of ergonomic analysis. In this step, an accurate "picture" of that work environment is developed. This involves defining the job in terms of duties and tasks, recording static workstation components and dynamic force demands, and often videotaping the performance of specific tasks.

RISK FACTOR IDENTIFICATION

A step-by-step comprehensive protocol is used to look at each body part and record specific types of risk factors. It is critical that nontechnical jargon be used to record and quantify each risk factor. This will facilitate better comprehension by all parties involved in acting upon this information.

INTERVENTION DISCOVERY

A creative critical thinking approach using a free-form model allows you to come up with unlimited possibilities for workplace improvement regardless of your level of experience in this area. A comprehensive approach ensures that all aspects of the work environment are considered. This includes workstation design, workstation modification, changes in process design, and worker-based programs.

INTERVENTION PLANNING

In this step, you consider the many requirements for implementing an intervention. This is an often missed critical step in an ergonomic process. Many otherwise valid and important interventions have failed because of a lack of realistic plans for the implementation for those interventions. An important aspect of this step is "intervention research." In many cases, it is not necessary or possible to design a new workstation or device. Individuals with limited experience can look for the type of product in an intervention database and request information directly.

IMPLEMENTING INTERVENTIONS

It is important to detail the steps in implementation of an intervention to facilitate line management and employee acceptance and compliance. This includes verifying the delivery, setup, orientation and training, and follow up.

PROJECT PROGRESS AND EFFECTIVENESS

Most interventions require ongoing support to modify or augment the intervention as needed. This can be assisted by ongoing monitoring of the impact on ergonomic criteria selected in initial steps and communicating with management.

RESPONSIBILITIES FOR AN EFFECTIVE ERGONOMIC PROCESS

Company name	Date
Facility location	

Who?	Responsibility
	I will meet with members of the facility management team to determine their perception of the need for ergonomic change.
	I will meet with members of the facility management team to see what information they would like to assist in their decision making.
	I will gather specific data to determine the jobs to be analyzed.
	I will analyze the data to determine the job to be analyzed.
	I will videotape the job being analyzed.
	I will measure and record the dynamic components of the workstation being analyzed.
	I will measure and record the static components of the workstation being analyzed.
	I will write a short narrative statement describing the job being studied.
	I will write a job duty list for the job being studied.
	I will write a task list for the job being studied.
	I will compile a list of ergonomic risk factors.
	I will compile a comprehensive list of interventions.
	I will consider all the interventions and determine which are possible.
	I will consider all the interventions and determine which are the most likely to succeed.
	I will look through advertisements and catalogs to determine if there are any appropriate interventions.
	I will visit other similar facilities to see what type of ergonomic interventions they are using.
	I will present a report to facility management that outlines suggested interventions.
	I will follow up with facility management to determine a start time for each intervention.
	I will research the various vendors to see what is available.
	I will analyze the information about the intervention and prepare a comparative analysis.

Who?	Responsibility
	I will contact other people who have used this intervention to determine how effective it has been for them.
	I will arrange for a small number of workers to test interventions.
	I will follow up with workers testing the interventions and document their feedback.
	I will write a report containing initial intervention feedback and provide it to facility management.
	I will follow up with facility management to be informed of a decision regarding overall implementation of the interventions.
	I will arrange the purchase of selected interventions.
	I will plan the orientation for each intervention.
	I will perform the orientation for each intervention and teach workers how to use it.
	Once the workers have started using an intervention, I will be available to answer their immediate questions about it.
	After the workers have used the intervention for a while, I will gather and correlate some measurement of success for that intervention.
	After the measurement of success data is gathered, I will compare it to the baseline data to assess the effectiveness.
	I will write a report to facility management regarding the measure of success for an intervention.
	I will follow up with facility management to find out about expanding, continuing, modifying, or abandoning interventions.
	Depending upon facility management's decision, I will communicate with the appropriate person (named above) to carry out that decision.
	I will follow up with the person responsible for carrying out management decisions regarding continuation, modification, or abandonment of interventions.
	I will report back to facility management regarding the progress of this intervention.
	Others

Worksheet 1–2 Ergonomic Analysis Format Question List: Instructions and Procedures

The purpose of this step is to identify the individual or individuals within an organization most likely to be able to make decisions regarding the implementation of interventions. You can significantly enhance your chances of creating a positive change in the work environment by tailoring your analysis to suit the decision maker in an organization. The benefit to this approach is that you're less likely to waste your time on unnecessary steps in an ergonomic analysis. It is also helpful to determine what the final outcome of an ergonomic intervention should be to be considered a success.

The tasks involved in this step include first determining who is the decision maker in an organization. Next, try to schedule a meeting with that individual. It is important to prepare for that meeting. This book provides you with several tools for achieving each of these tasks as well as for recording your impressions based on your meeting.

PROCEDURES FOR IDENTIFYING THE DECISION MAKERS IN AN ORGANIZATION

1. Consult with initial contact or others in an organization.
2. Select appropriate questions from the Ergonomic Analysis Format Question List.
3. Add any necessary questions and delete the inappropriate ones.
4. Schedule the first meeting.
5. Rehearse the questions you will ask.
6. If you have any, bring samples of different report formats with you to show the various options.
7. Attend the meeting. Make sure you know how much time you have.
8. Immediately after the meeting, review the questions and responses and record any additional impressions.

ERGONOMIC ANALYSIS FORMAT QUESTION LIST

Introduction

I am going to be performing ergonomic analyses and would like to discuss your expectations for this project. By doing this, I will be able to focus my efforts and make the most of the time that is available. My goal is to provide you with the necessary information to allow you to make an intelligent decision regarding the need for any changes.

Decision-Making Perspective

	How long have you been with the company?
	Have you always been in a management position?
	Are there opportunities for advancement?
	What would you like to see happen while you are in your current position?
	Have you found that your accomplishments seem to "get lost" after you leave whereas any mistakes seem to follow you wherever you go?
	What type of changes have you been responsible for in your current position?
	What kind of criteria are used for making decisions within the organization?
	Why do you work only one shift per day?
	How successful is your business currently?
	Will the current political climate affect your business in any way?
	Are you planning to hire many people in the next few years?

Ergonomic Attitude Questions

	Have other ergonomic evaluations been performed for your company?
	Have you initiated any changes? How effective have they been?
	How serious do you believe the need for workplace ergonomic modification is?
	How urgent do you believe the need for workplace ergonomic modification is?
	Do you think workstation changes are necessary or do you think the same results can be achieved through changes in worker performance?
	Which do you think is the greatest driving force for this project—Occupational Safety and Health Administration citation for ergonomic violations, incidence of injury, or workers' compensation costs?
	How soon do you think you will be able to initiate a change?
	Where do you think the greatest need for an ergonomic analysis is?
	Are there any plans under way to eliminate or dramatically redesign any workstations?
	If training is necessary for employees, how much is available?
	If training is necessary for managers and supervisors, how much is available?
	If capital expense is necessary, will it be available this year or will it be part of next year's budget?

Ergonomic Report Questions

	What type of information can I provide that will help you the most?
	What type of information can I provide to help you assess the need for initiating any changes?
	What type of information can I provide to help you select a specific intervention?
	How soon would you like to see the report?
	What format would you like the report to be in?
	Do you want to see the whole report, or would you rather have a short abstract of the report?

Success Measurement

	What would you like to see happen in the targeted area?
	What would have to happen for this project to be considered an overwhelming success?
	How will you measure the effectiveness of any intervention that is implemented?

Additional Questions

Final Report Format Questions

What Format Should the Final Report Take?

	Live
	Live with videotape
	Live with slides
	Written
	Written with still pictures

What Information Should Be in the Final Report?

	Baseline criteria
	Explanation of baseline criteria
	Narrative job description
	Duty list
	Task breakdown
	Static workstation diagram
	Dynamic forces record
	Videotape
	Ergonomic risk factor list
	Comprehensive list of interventions
	Summarized list of interventions
	Plan for implementation of interventions
	Cost of proposed interventions
	Summary and recommendation (two to three pages)
	Report abstract (one page)

Worksheet 2-1 Criteria Selection and Baseline Statement: Instructions and Procedures

The purpose of this step is to determine the most appropriate data to be used, first as criteria for measuring the need for ergonomic intervention and then as a baseline for gauging the effectiveness of modifications in that work environment.

Injury incidence rate can help reflect the comfort and long-term employee well-being. This can be found or calculated by looking at an Occupational Safety and Health Administration (OSHA) 200 log or an in-house medical department visit log. The method for comparing different areas or setting up a baseline for long-term comparison is to calculate the incidence rate. This rate is accepted as a standard and will allow you to compare your facility to other facilities or industries.

The criteria selection and baseline statement provides a method for clearly stating who is currently responsible for recording the information, who will be responsible for collecting it, how often they will collect it, and whatever that measurement is at this point. This information will eventually be included in the introductory statement of your final ergonomic analysis.

USING THE FORM FOR CALCULATING INCIDENCE RATE

If the personnel department has the total number of hours employees work in a year, go to step 4.

1. Write down the total number of employees in a department or facility.
2. Write down the number of hours each employee works per week.
3. Write down the number of weeks worked in a year.
4. Multiply the numbers in steps 1, 2, and 3 together. This is the total number of hours worked in a year. Enter this figure in number of hours worked column.
5. Write down the total number of injuries (or events) that occurred in a year in the number of injuries column. Perform arithmetic as indicated.

COMPLETING THE CRITERIA SELECTION AND BASELINE STATEMENT

1. Enter identification information regarding the workstation, department, and facility.
2. Enter the three or more criteria that will be used to demonstrate the need for

ergonomic analysis and act as a baseline for periodic measurement of the effectiveness of intervention.
3. Enter the individuals who will be responsible for recording and collecting the data selected.

FORM FOR CALCULATING INCIDENCE RATE

$$\text{Incidence rate} = \frac{\text{Number of injuries or illnesses in a year} \times 200{,}000}{\text{Actual number of hours worked by all employees in a year}}$$

Department	Number of Injuries				Number of Hours Worked	Incidence Rate
		× 200,000 =		÷		
		× 200,000 =		÷		
		× 200,000 =		÷		
		× 200,000 =		÷		
		× 200,000 =		÷		
		× 200,000 =		÷		
		× 200,000 =		÷		
		× 200,000 =		÷		
		× 200,000 =		÷		
		× 200,000 =		÷		
		× 200,000 =		÷		
		× 200,000 =		÷		
		× 200,000 =		÷		
		× 200,000 =		÷		
		× 200,000 =		÷		
		× 200,000 =		÷		
		× 200,000 =		÷		
		× 200,000 =		÷		

CRITERIA SELECTION AND BASELINE STATEMENT

The following criteria have been selected to act as a baseline for the ergonomic analysis of the

_____ workstation located in the

_____ department of

_____ facility.

These measurements were selected based on objectivity, availability, and consistent tracking procedures.

Criterion 1 _____

Criterion 2 _____

Criterion 3 _____

Criterion 1 is recorded by _____

and will be collected by _____

every _____ years/months.

It is currently _____ for this (facility/department/job title).

Criterion 2 is recorded by _____

and will be collected by _____

every _____ years/months.

It is currently _____ for this (facility/department/job title).

Criterion 3 is recorded by _____

and will be collected by _____

every _____ years/months.

It is currently _____ for this (facility/department/job title).

Worksheet 2-2 Ergonomic Need Criteria Comparison Chart: Instructions and Procedures

The purpose of this step is to determine the most appropriate data to be used, first as criteria for measuring the need for ergonomic intervention and then as a baseline for gauging the effectiveness of modifications in that work environment.

You may choose to look at productivity variances. These variances may be productivity that is either higher than expected or lower than expected in a particular area. Quality variances can also be an indicator of ergonomic need. This can be assessed by checking reject or rework records. Finally, injury incidence rate can reflect long-term employee well-being and comfort. This can be found or calculated by looking at an OSHA 200 log or an in-house medical department visit log.

There are other, more indirect measures of a need for ergonomic intervention in an area. Turnover and absenteeism may point to a need for ergonomic intervention but are often more indicative of the relationship between the workers and management.

COMPLETING THE ERGONOMIC CRITERIA COMPARISON

1. Select criteria to be investigated.
2. Compile data to be assessed.
3. Fill in Ergonomic Need Criteria Comparison Chart.
4. Determine jobs or areas with greatest need for ergonomic analysis and fill in Criteria Selection and Baseline Statement in Worksheet 2-1.

ERGONOMIC CRITERIA COMPARISON CHART LEGEND

Number of injuries—Number of injuries in the department listed.

Percent of injuries—Percent of injuries occurring in a department relative to total injuries occurring in the facility ([department injuries/total injuries] × 100).

Productivity variance—Percent of target productivity goals (100% means met target, less than 100% means produced less, more than 100% means exceeded production goals).

Rework/reject variance—A simplified measure of quality, reflecting how many products need to be repaired or replaced. (100% means met target, less than 100% means poor performance, more than 100% means better than expected performance [fewer rejects]).

Number of lost days—Total number of lost days as a result of an injury.

Turnover per 100 workers—Number of people that transferred or resigned from that department. This number is reduced or expanded to reflect a department with 100 workers in it.

Absent per 100 workers—Number of days that people from the department were absent in the period being studied. This number is reduced or expanded to reflect a department with 100 workers in it.

Comfort-level survey—Empirical average of comfort or discomfort based on anonymous survey (0 means no discomfort, 1 means discomfort rarely, 2 means discomfort more often).

Number of cumulative trauma disorder cases—The number of reported cumulative trauma disorder cases in the department during time studied.

ERGONOMIC NEED CRITERIA COMPARISON CHART

	Direct Measurement				Indirect Measurement			
Department	Number of injuries	Percent of injuries	Productivity variance	Rework/ reject variance	Lost days	Turnover per 100	Absent per 100	Comfort-level survey	Number of cumulative trauma disorder cases

Worksheet 2–3 Subjective Comfort-Level Survey: Instructions and Procedures

The purpose of this step is to gather and analyze the subjective comfort level in a particular job title or area and convert it to a measurable objective unit.

PROCEDURES FOR GATHERING DATA

1. Choose the body parts that you wish to monitor or assess. Make a copy of the Subjective Comfort-Level Survey in this worksheet. Cross out any sections for which you do not wish to collect data.
2. Make copies of the form that workers will fill out.
3. Include survey as a part of a regular employee meeting or as a special event.
4. Explain that the results are anonymous and that this is part of an overall effort to improve their job.
5. Provide the instructions as follows:

    Please complete the following survey by circling the answer that best describes your experience. This form will be used to measure the effectiveness of our efforts to increase the comfort and safety of your job. This information is completely anonymous. Please be totally honest in your assessment.

6. Collect and score the surveys using the procedure outlined below.

PROCEDURES FOR PROCESSING DATA

1. Add up the responses to each question. The form was designed to allow you to overlap the left edges of the completed surveys and then add up the numbers across for the "mild" responses. The surveys can then be overlapped in the opposite direction to add up the "strong" responses.
2. Divide the sum for each body part by the total number of respondents. This will give you the average frequency that individuals experience either mild or strong discomfort in the body part in question.
3. The frequency for mild and for strong discomfort cannot be combined.
4. The frequency for mild discomfort at different body parts *can* be combined to get a general idea regarding comfort level.

SUBJECTIVE COMFORT-LEVEL SURVEY

Please complete the following by writing the number that best describes your experience.

1 = Rarely or never
2 = One or two days per month
3 = Several days per month
4 = Every day or almost every day

	I experience MILD discomfort in my FEET.	I experience STRONG discomfort in my FEET.	
	I experience MILD discomfort in my KNEES.	I experience STRONG discomfort in my KNEES.	
	I experience MILD discomfort in my LOW BACK.	I experience STRONG discomfort in my LOW BACK.	
	I experience MILD discomfort in my BACK/NECK.	I experience STRONG discomfort in my BACK/NECK.	
	I experience MILD discomfort in my SHOULDER.	I experience STRONG discomfort in my SHOULDER.	
	I experience MILD discomfort in my HAND/WRIST.	I experience STRONG discomfort in my HAND/WRIST.	
	I experience MILD discomfort in my _____.	I experience STRONG discomfort in my _____.	

Worksheet 3-1 Introduction and Job Description: Instructions and Procedures

This worksheet will assist in organizing the way you gather information and record observations as you perform your ergonomic analyses. Using the worksheet can either facilitate writing a narrative ergonomic analysis report or serve as a stand-alone record of your work.

BACKGROUND INFORMATION AND WORKER IMPRESSIONS

1. Spend time talking with human resources, line management, and current and past workers in the target area.
2. Use the following pages on job background information and worker impressions to help structure and record these discussions.

Use the information gathered from these interviews to complete the job description and duty list page.

JOB DESCRIPTION AND DUTY LIST

Identifying Information

1. Enter company name.
2. Enter job title and number, if available.
3. Enter name of person or group performing the analysis and date.

Job Summary

1. Review any available job description and interview on-line supervisors and hourly workers.
2. Based on information gathered, enter a brief narrative that reflects the primary duties and responsibilities of the job being analyzed.

Job Exposure Calculation

1. Enter "on duty" time.
2. Enter number of minutes of the first coffee break of the day.
3. Enter the total number of minutes for all meal breaks taken during a shift.
4. Enter the number of minutes of the second coffee break of the day.
5. Enter any additional down time.
6. Use the workspace on the right side of the paper to calculate job exposure.

Duty List

1. Compile a list of duties based on information gathered from client in the form of written job description or on-site verbal input.
2. Calculate the percentage of job exposure for all duties listed using the following formula:

Duty percentage = (Duty exposure/Total job exposure) × 100

JOB BACKGROUND INFORMATION

Company name					
Company contact					
Address					
Phone/fax					
Reason for analysis					
Information provided by					
Work week		hours per day		days/week	
Breaks	Frequency	per day		Duration	minutes/break
Meals	Frequency	per day		Duration	minutes/break
Rotation	Yes/No	**How often?**		**Number of stations**	
Rotation details					
Regulatory hours of service restrictions?					
Paid by	Hour	Day	Piece	Quota	Bonus
Production rate required?	Yes/No				
Multiple workstations?	Yes/No		**Workload assigned at shift start?**	Yes/No	
Teamwork?	Works alone	With others	Necessary part of crew/team		
Supervised directly?	Yes/No		**Supervisory duties?**	Yes/No	
Indoors/outdoors					
Climate hazards					
Types of equipment/vehicles used					
Protective equipment required					
Dress code/uniform?	Yes/No				
Training necessary?	Yes/No				
Job transfers	Possible/Required		**Promotions**	Possible/Required	
Union representation?	Yes/No		**Part-time available?**	Yes/No	
Other					

WORKER IMPRESSIONS

How long have you been at this job?
What do you like best about this job?
What do you do? (What are job duties?)
What is the most difficult part of the job?
Is there any uncomfortable part of the job?
Do you have any suggestions for changing the process?
Do you have any suggestions for changing the workstation?
Additional comments

JOB DESCRIPTION AND DUTY LIST

Company name		
Job title		Job number
Analyst name		Date

Job Summary

Job Exposure Calculation

On-duty time	
Number of minutes for first break	
Number of minutes for lunch	
Number of minutes for second break	
Any other down time	
Off-duty time	

Duty List

Duty number		Hours per day	Percentage

Worksheet 3-2 Workstation Measurement: Instructions and Procedures

The purpose of this step is to objectively and carefully record data about the physical description of the work environment. This includes work heights, reach distances, walking distances, overhead or under desk clearances, and any other pertinent data relative to the physical environment. This information is recorded to establish a baseline for modification or for comparison with other workstations in the future.

1. Take a picture of a workstation from the side. For the purpose of consistency in all your pictures, try to shoot the picture at the operator's eye level.
2. Mount the picture on the worksheet with two-sided tape or glue.
3. Find a spot within your picture from which you will be able to measure all the distances clearly. Use a permanent ink marker to draw a line from the reference point on the picture directly out to the side and directly down. Mark it "zero." This will be your reference point.
4. On the picture, draw lines out from each of the distances to be measured. Try to make each line a little longer as you get further from the reference point.
5. Use the worksheet as a guide to measure the various distances on the workstation. Enter the data at the ends of the appropriate lines drawn on the worksheet.
6. Move to a position about 90 degrees (to either in front or in back) from the original and take another picture.
7. Repeat steps 2 through 5.
8. If necessary, move another 90 degrees and take a picture from the opposite side. Remember to keep the camera at the operator's eye level.
9. If no camera is available, sketch the workstation.

WORKSTATION MEASUREMENT RECORD

Workstation	Date

View	☐ Front	☐ Back	☐ Side	☐ Overhead	☐ Other

Unit	☐ Inches	☐ Feet	☐ Centimeters	☐ Meters	☐ Other

PASTE PHOTOGRAPH HERE

Workstation				Date	

View	☐ Front	☐ Back	☐ Side	☐ Overhead	☐ Other

Unit	☐ Inches	☐ Feet	☐ Centimeters	☐ Meters	☐ Other

PASTE PHOTOGRAPH HERE

Worksheet 3-3 Dynamic Force Measurement: Instructions and Procedures

The purpose of this step is to compile a list of statements that describe the forces that must be overcome to move or control objects in the work environment. This information can be helpful in describing the specific physical demands of a job as well as providing a baseline for measuring the effectiveness of workstation improvements.

Based on your job summary statement and direct observation, you will measure the dynamic forces associated with the performance of this job. This includes weights, pushing and pulling forces, and torque necessary to perform a task.

MEASURING PUSHING AND PULLING FORCES USING A FORCE GAUGE DYNAMOMETER

1. Try to attach the force gauge in a way that closely simulates the way the worker's hand or other body part normally makes contact with the object.
2. Use devices such as vice grips, wood clamps, paper clips, chains, or cables to attach the objects being tested to the force gauge dynamometer. A small section of decorative fish net and a ball of strong twine seem to meet most normal coupling needs.
3. Select the most appropriate form to record your results by considering the degree of detail that will be required to defend or explain your data.
4. Use a smooth and safe application of force. Record the force necessary to initiate movement as well as the force necessary to maintain movement. Be aware of uphill and downhill grades, rugs, or other resistive flooring.

DETAILED MEASUREMENT WORKSHEET

Workstation	Measured By	Date

Trial 1	Trial 2	Trial 3	Trial 4	Trial 5	Trial 6	Average	Units

Method of attachment (if any)
Additional comments

Force (no.)	Units (lb, kg)	Are Required To Push, Pull, Lift, or Hold (select one)	Task Description

Draw diagram of measurement technique used or paste picture in space below.

Appendix G 143

SUMMARY RECORD

Workstation	Measured By	Date

Force (no.)	Units (lb, kg)	Are Required To Push, Pull, Lift, or Hold (select one)	Task Description

Force (no.)	Units (lb, kg)	Are Required To Push, Pull, Lift, or Hold (select one)	Task Description

Force (no.)	Units (lb, kg)	Are Required To Push, Pull, Lift, or Hold (select one)	Task Description

Force (no.)	Units (lb, kg)	Are Required To Push, Pull, Lift, or Hold (select one)	Task Description

Worksheet 3–4 Videotape Recording Log: Instructions and Procedures

This worksheet provides a model for video recording a target job and then setting up a written log. The log will assist you in locating specific sections of tape during your analysis. This videotape will later allow you to closely examine and analyze not only what a person does but also whether any conditions or practices that potentially can harm him or her are being performed. The videotape log provides a reference for locating the sections of videotape of specific jobs, duties, or tasks. This is best done as soon as possible after shooting the videotape. It provides a place for entering the date, identification of the information from the tape, and a record of who shot it and location.

SHOOTING VIDEOTAPE

1. Start with an establishing shot as far away from an individual as possible. With the lens zoomed out as far as it will go, show his or her whole body, the workstation, and the surrounding area.
2. Zoom in until you see the entire body. Shoot an entire task cycle or entire repetition of activities at this distance.
3. Zoom in closer and shoot just the hands, the feet, or the upper body performing tasks considered important.
4. Change the angle at which you are shooting to reveal a different view and repeat the process.

SETTING UP A VIDEOTAPE LOG

1. Rewind the tape to the beginning using a tape player that you will have access to on a regular basis.
2. Set the counter to zero and then start to watch the tape.
3. Whenever anything significant changes, for instance the distance that you are from the individual performing the task or the task that he or she is performing, stop the tape player, record the counter number, and then start the tape again. Do not reset the counter to zero.

HINTS

1. Take advantage of the camera's capability to display date and time, if available.
2. Use the screen counter if available.

3. Label the tape before you use it.
4. Always carry spare batteries and tape with you.
5. Never shoot more than 20 minutes on one tape.
6. Turn the camera off or cover the lens with your hand to break up segments.
7. Shoot from the hip.
8. Turn on the camera to record conversation but point the camera at the ground or keep the lens cap on.
9. When you are finished with a tape cassette, immediately break out the tab to prevent recording over 10. Do not preview the tape.
10. If the camera has a red light that indicates it is on, cover the light with dark tape or disable it.
11. If workers are changing their behavior whenever you turn the camera on, pretend to shoot and see if they get used to it.
12. Few workstations lend themselves to the previous procedures.

VIDEOTAPE RECORDING LOG

Date			Tape ID Number	Recorded By	
Location					
Start	End	Date/Time	Description		

Worksheet 3–5 Job Task Breakdown: Instructions and Procedures

The Job Task Breakdown is a detailed list of activities that each task involves. In addition, each task can be further described in terms of exposure, rate, and quantity. The exposure is how long a task is performed. Rate is the speed at which the task is performed. Quantity is the number of times a task is performed.

This information can provide a detailed framework upon which to base subsequent analysis steps. It will provide a method for comparison among the various activities and facilitate the quantification of ergonomic risk factors.

TASK BREAKDOWN PROCEDURES

1. Enter duty number and description in the duty description box.
2. Directly observe or review videotape of a job being performed.
3. Enter the tasks that are involved in performing the duty in the task number/description area in chronological order.
4. Determine the unit of completion for each task and enter.
5. Enter the total number of hours it takes for this task to be performed.
6a. If total number of daily units of completion is available, enter that number in the quantity column. Determine the rate by dividing the quantity by the exposure and enter in rate.
6b. If total number of daily units of completion is not available, but the rate is available, take the number of hours that this task is performed and multiply by the rate to determine the quantity. If total number of daily units of completion is not available and the rate is not available, determine the rate by observing the worker perform the task for a specified period. Convert that rate to an hourly rate.
7. Measure the amount of time it takes to perform one unit of completion.
8. Enter performance time in the performance time box.

TASK CYCLE BREAKDOWN

1. List the tasks in the order that they are performed.
2. Enter the first task in the first box on the left. Enter the next task on the next line but indent it one box to the right.
3. Continue to add tasks, each time indenting one box. If a task repeats, enter it on

its own line but start it in the same column as the first time you wrote it on the worksheet.
4. Continue until the first task entered on the form repeats. When that happens, you have defined the beginning and end of your task cycle.

TASK BREAKDOWN WORKSHEET

Duty number/description		Task number/description		
Units of completion	Performance time	Exposure	Rate	Quantity
		Hours per day	Units per hour	Units per day

Duty number/description		Task number/description		
Units of completion	Performance time	Exposure	Rate	Quantity
		Hours per day	Units per hour	Units per day

Duty number/description		Task number/description		
Units of completion	Performance time	Exposure	Rate	Quantity
		Hours per day	Units per hour	Units per day

Duty number/description		Task number/description		
Units of completion	Performance time	Exposure	Rate	Quantity
		Hours per day	Units per hour	Units per day

Duty number/description		Task number/description		
Units of completion	Performance time	Exposure	Rate	Quantity
		Hours per day	Units per hour	Units per day

TASK CYCLE BREAKDOWN WORKSHEET

Worksheet 4-1 Identification of Risk Factors: Instructions and Procedures

The purpose of this step is to consistently and comprehensively identify ergonomic risk factors in a work environment. An ergonomic risk factor is a condition or practice that can act as an obstacle to productivity, a challenge to consistent quality, or a threat to worker comfort, safety, and long-term well-being.

PROCEDURES FOR IDENTIFYING ERGONOMIC RISK FACTORS

1. Decide whether you will be observing each task or if you will simply be watching someone performing several tasks that make up a duty and identifying risk factors.
2. Observe the individual performing his or her job either in person or on videotape. Watch the entire process to gain familiarity.
3. Starting at a person's feet and working your way up through the knees, thighs, low back, shoulders, arms, and so on, continue to observe any awkward range positions the worker assumes.
4. Record the body part, position, and frequency on the form.
5. Starting at a person's feet and working your way up various body parts, observe and record any unsupported positions the worker maintains.
6. Enter the body part, position, and duration on the form.
7. Continue to observe all body parts consistently to observe and record any forceful exertion the worker applies.
8. Enter the type of forceful exertion, the amount of force to be overcome, and frequency.
9. Continue to observe and record any body parts exposed to environmental conditions. It is suggested that at first you focus on individual risk factors and observe the entire body rather than look at the body one part at a time observing various risk factors.
10. Once you have completed this body-part-by-body-part examination, then look for any signs of excessive physiologic demand. Ideally, a task-by-task, body-part-by-body-part risk identification procedure will result in the most comprehensive list of risk factors. It will also result in an extensive list of risk factors.

A list of types of risk factors and suggestions for recording shortcuts follows.

- An awkward range position occurs when a person moves a body part as far as it will go or close to this position.
    AR = Assumes an awkward range position
- An unsupported position occurs when an individual holds a body part without moving it or resting it on anything for a period.
    UP = Maintains an unsupported position
- A forceful exertion occurs when an individual moves a body part against resistance or maintains a body part in a static position against resistance.
    FEL = Applies a lifting forceful exertion
    FES = Applies a static forceful exertion
    FE–PS = Applies a pushing forceful exertion
    FE–PL = Applies a pulling forceful exertion
    FE–TW = Applies a twisting forceful exertion
- An environmental condition is an element of the physical surroundings that may result in discomfort or interrupted productivity.
    Body part is exposed to:
    EN–HE = heat or hard surface
    EN–C = cold or cold surface
    EN–HD = hard or sharp surface
    EN–VI = vibration
    EN–OT = other
- Excessive energy demand is a requirement for physical exertion that by nature of its excess can lead to discomfort or interrupt productivity.
    Excessive energy demand as indicated by:
    EX–SW = profuse sweating
    EX–SP = inability to speak due to heavy breathing
    EX–RE = reaching for support
    EX-RU = rubbing body part

The following is a list of body positions using layperson's terms. Using consistent terms to describe risk factors and body positions is essential for consistent and comprehensive risk identification.

- Foot and Knee
    Points/pushes up on toes

Pulls end of foot up
  Turns foot in
  Turns foot out
  Bends knee
  Straightens knee
- Hip
  Pulls knee up toward body
  Moves whole leg back at hip
  Moves leg out to side
  Moves leg across in front
  Turns leg in at hip
  Turns leg out at hip
- Low Back
  Bends at waist
  Bends backward at waist
  Twists at waist
  Bends to the side
- Neck
  Bends neck down or forward
  Bends neck up or back
  Moves head to side
  Turns head
- Shoulders
  Moves arm overhead
  Moves arm back or behind
  Moves arm to side and overhead
  Moves arm across body
  Slumps shoulder forward
  Pulls shoulder backward
  Turns arm in at shoulder
  Turns arm out at shoulder
- Elbow
  Bends elbow
  Bends elbow backward
- Wrist
  Bends wrist down
  Bends wrist backward
  Bends wrist (to pinky side)

    Bends wrist (to thumb side)
    Turns hand over (down or in)
    Turns hand over (up or out)
- Hands/Fingers
    Closes hand into fist
    Bends fingers backward
    Spreads fingers apart
    Bends fingers down/in

RISK IDENTIFICATION EXAMPLES

Job title/duty/task	Chopping onions				
Type	Body part	Position	Amount of force	Environmental conditions	Frequency/ duration
Awkward range	Wrist	Bends to pinky	NA	NA	150/onion
Comments: Worker chops about 30 onions twice a day.					

Job title/duty/task	Chopping onions				
Type	Body part	Position	Amount of force	Environmental conditions	Frequency/ duration
Unsupported	Low back	Forward bend	NA	NA	30 minutes
Comments: Worker chops onions twice a day.					

Job title/duty/task	Puts boxes into walk-in refrigerator				
Type	Body part	Position	Amount of force	Environmental conditions	Frequency/ duration
Forceful exertion—Lifting	Shoulders	NA	35 lb	NA	15 boxes
Comments: Worker puts away boxes of food as they are delivered. Approximately 15 boxes are delivered each day.					

Job title/duty/task	Puts boxes into walk-in refrigerator				
Type	Body part	Position	Amount of force	Environmental conditions	Frequency/ duration
EN—Cold	Hands	NA	NA	Cold	15 boxes
Comments: Worker puts away boxes of food as they are delivered. Approximately 15 boxes are delivered each day.					

Job title/duty/task	Puts boxes into walk-in refrigerator				
Type	Body part	Position	Amount of force	Environmental conditions	Frequency/ duration
EX—SP					
Comments: Worker is unable to speak normally when putting boxes into refrigerator.					

FORM FOR RISK IDENTIFICATION

Job title/duty/task					
Type	Body part	Position	Amount of force	Environmental conditions	Frequency/ duration
Comments:					

Job title/duty/task					
Type	Body part	Position	Amount of force	Environmental conditions	Frequency/ duration
Comments:					

Job title/duty/task					
Type	Body part	Position	Amount of force	Environmental conditions	Frequency/ duration
Comments:					

Job title/duty/task					
Type	Body part	Position	Amount of force	Environmental conditions	Frequency/ duration
Comments:					

Job title/duty/task					
Type	Body part	Position	Amount of force	Environmental conditions	Frequency/ duration
Comments:					

Worksheet 5-1 Interventions To Create a Safe and Effective Work Environment: Instructions and Procedures

The purpose of this worksheet is to help you compose a comprehensive list of possible interventions to create the safest, most comfortable, and most effective work environment possible. Try to be creative and nonjudgmental about the ideas you come up with. The most preposterous idea may stimulate a more realistic intervention later. Also, avoid looking for the one "right" answer. There seldom is just one.

To assist in your consideration of the many options available, it is helpful to focus on specific types of interventions. This ensures a comprehensive thought process rather than just getting stuck on the last great idea you read about or saw advertised in a magazine. The categories for your consideration are worker-based interventions, process-based interventions, and workstation-based interventions. Each of these can be broken into more specific subcategories. They are listed and briefly described in this worksheet.

PROCEDURES

1. Enter all possible and even impossible interventions in the appropriate boxes as prompted throughout the worksheet.
2. After completing the worksheet, look over your ideas and select two or three that seem to be the best. Make note of them on the last page.
3. Put this away for a couple of days and as other ideas come to mind, enter them on the worksheet. It may become necessary to add pages.

WORKER-BASED INTERVENTIONS

Training

Behavior to be adopted or continued

_____
_____
_____

Training content (Least stressful work practices? Equipment adjustment? Maintenance criteria? Exercises?)

_____
_____
_____
_____

Training format (Live? Video? Poster? Handout?)

_____
_____

Training frequency (Orientation? Yearly? After injury? Quarterly?)

_____

Exercises

Types of exercise (Warm-up? Strengthening? Stretching?)

_____
_____
_____

Frequency of exercise (Once per day? Regular group exercise? Independent?)

_____
_____
_____

Protective Equipment

Type of personal protective equipment (Shoe insoles? Gloves? Backbelts? Wrist splints? Elbow pads?)

_____
_____
_____
_____

Rotation

Rotate workers between different jobs? Allow worker to vary tasks?

_____
_____
_____

How often? (Every other day? Every few hours? Workers' choice?)

_____

Procedure Mandate

Procedures to be mandated (Rules? Policies? Protocols?)

_____
_____

Method of communication (Orientation? Employee handbook? Meetings?)

_____

Method of compliance monitoring (Supervisor? Video? Electronic device?)

_____
_____

Penalty for noncompliance

_____

PROCESS-BASED INTERVENTIONS

Input

What is the input? (Item or items to be processed, moved, or acted upon in any way.)
_____
_____

How can it be modified? (Position of the input? Rate at which it is presented? Size? Shape? Method of packaging? Weight? Temperature?)
_____
_____
_____
_____

Output

What is the output? (Final condition of that input as a result of the actions taken by the worker.)
_____
_____

How can it be modified? (Number of steps performed? Final position of the product? Speed of output?)
_____
_____
_____
_____

WORKSTATION-BASED INTERVENTIONS

Workstation Adjustment

What can be adjusted on the workstation?
_____
_____

In what direction? (Height? Depth? Distance? Rotation? Angle?)

_____

_____

_____

_____

_____

Help To Move

What is being moved?

_____

How can it be assisted?

_____

_____

_____

_____

Help Hold in Place

What is being held in place?

_____

How can it be assisted?

_____

_____

_____

_____

Decrease Repetitions

What task is being performed?

_____

_____

What can be done to decrease repetitions while maintaining productivity?

_____
_____
_____
_____

Support Operator

What body part is unsupported?

_____

How can it be supported?

_____
_____
_____
_____

Facilitate Access

What physical items are being accessed or retrieved?

_____
_____

How can access port be turned, tilted, or enlarged?

_____
_____
_____
_____

Modify Environment

What are environmental conditions to be modified?

_____
_____

How can they be modified, altered, or controlled?

_____
_____
_____
_____

Facilitate Information Input

What information is being seen, heard, read, or received in some other way?

_____

How can information input be made bigger, louder, clearer, or less obstructed?

_____
_____
_____
_____

Modify Controls

What are the controls? (On/off switch? Speed variation? Steering?)

_____
_____

What alternatives would make using them easier or more comfortable?

_____
_____

Maintenance Programs

What equipment parts or tools need to be maintained to make job easier?

_____
_____

How can regular maintenance help? What criteria can be used to request maintenance?

___

___

Tool Options

What activities could be easier with a tool?

___

What kind of tool?

___

___

___

Unclassified and Other Great Ideas

Is there anything else that comes to mind?

___

___

___

INTERVENTION DISCOVERY GUIDE

The following examples may prove helpful if you cannot think of any on your own.

Worker-Based Ideas

Education Program Format Options

- ☐ Initiate a live presentation training program.
- ☐ Use a videotape education and training program.
- ☐ Initiate an interactive computer video training program.
- ☐ Use a generic slide presentation.
- ☐ Train in-house personnel to perform ongoing training.
- ☐ Distribute literature and posters.
- ☐ Provide five-minute per day training programs.
- ☐ Have a customized videotape, slide, or live presentation produced.

Education Program Content Options

- ☐ Least stressful work practices (lifting, job-specific procedures)
- ☐ Potentially harmful home activities
- ☐ How to adjust the equipment or seating
- ☐ How to adjust the workstation
- ☐ How to maintain tools
- ☐ How to recognize the need for maintenance
- ☐ Mandated procedures
- ☐ Need for proper-fitting clothes
- ☐ Need for proper-fitting personal protective equipment
- ☐ Principles of ergonomics
- ☐ Performance methods
- ☐ Postures to avoid

Exercise Program Content Options

- ☐ Initiate an exercise program for upper extremity flexibility and strength maintenance.
- ☐ Initiate an exercise program for low back flexibility and strength maintenance.
- ☐ Initiate an exercise program for overall flexibility and strength maintenance.
- ☐ Initiate an exercise program for overall cardiovascular conditioning.

Exercise Program Format Options

Exercises will be performed

- ☐ at beginning of shift.
- ☐ on site immediately after meal break.
- ☐ on site voluntarily before the start of shift.
- ☐ at home.
- ☐ on site at beginning of shift and after meal break.

☐ on site at end of workday.
☐ on site independently throughout the day at unspecified intervals.
☐ on site as a group throughout the day at unspecified intervals.
☐ on site as a group throughout the day at specified intervals.
☐ in off-site health club.

Mandate Procedures

☐ Mandate specific procedures and hold each worker accountable for group performance.
☐ Mandate specific procedures and hold on-line management accountable for group performance.
☐ Mandate specific procedures and set up demerit system that leads to suspension or layoff.
☐ Mandate specific procedures and reward compliance with incentive programs.

Personal Protective Equipment

☐ Have workers wear wrist splints.
☐ Have workers wear hard back braces.
☐ Have workers wear lift belts.
☐ Have workers wear shock-absorbing inserts in shoes.
☐ Have workers wear insulated gloves.
☐ Have workers wear rubber gloves.
☐ Have workers wear shock-absorbing gloves.

Rotation

☐ Initiate a light duty/heavy duty rotation program.
☐ Initiate work/rest cycles.
☐ Initiate weekly, daily, or daily periodic rotation to jobs using different muscle groups.
☐ Allow workers to plan out their workday.
☐ Initiate valid light duty/break-in period for transferees.

Process-Based Ideas

Input Options

☐ Increase/decrease the size of the container/item.
☐ Modify the position of the container/item.
☐ Increase/decrease the weight of the container/item.
☐ Change the shape of the input.
☐ Have materials delivered to a different area.
☐ Have the supplier perform part of the task.
☐ Change the composition of raw materials.
☐ Increase/decrease the number of different pieces to be handled.
☐ Alter the properties of the materials (malleability, heat conductivity, etc.).
☐ Alter environmental conditions that affect material properties.
☐ Change the container or packaging that contains the materials.
☐ Change the informational delivery process.

- ☐ Allow personal control over the input rate.
- ☐ Increase specifications for more rigid standards for input materials.

Output Options

- ☐ Change the final placement location or position.
- ☐ Change the order of tasks.
- ☐ Change the output rate.
- ☐ Increase or decrease the number of steps each worker performs.
- ☐ Have worker pack or put away whatever he or she processes.
- ☐ Make the output rate variable.
- ☐ Do not act upon the input (ship as is).
- ☐ Ensure that current specifications are necessary.
- ☐ Stop using the currently used storage space.

Workstation-Based Ideas

Workstation Adjustment

- ☐ Adjust chair height, angle, depth, or armrest.
- ☐ Replace current chairs with adjustable models.
- ☐ Place the worker on a platform.
- ☐ Place worker in trough or sunken floor.
- ☐ Use pallets to raise worker.
- ☐ Provide adjustable floor height.
- ☐ Provide sit/stand chair.
- ☐ Use fixed or adjustable scaffolding.
- ☐ Place a cushion on existing seat.
- ☐ Place something on top of workstation.
- ☐ Place something under workstation.
- ☐ Attach workstation to wall.
- ☐ Cut two or four legs or bottom of workstation.
- ☐ Sink floor or place workstation in trough.
- ☐ Split work-surface heights.
- ☐ Use a spring-loaded double bottom cart.
- ☐ Place work surface on a cart.
- ☐ Install adjustable workstation.
- ☐ Install Lazy Susan (turntable).
- ☐ Cut out work surface.
- ☐ Change to bigger or smaller wheels.

Movement Assistance

- ☐ Add a motor drive to carts.
- ☐ Bevel work surface edges.
- ☐ Change or install handles on carts.
- ☐ Install a feeder mechanism.

- ☐ Install cart-loading mechanism.
- ☐ Install Lazy Susan.
- ☐ Install or change wheels.
- ☐ Install overhead swinging arm.
- ☐ Install permanent handles.
- ☐ Install temporary or special handles.
- ☐ Install overhead gantry, boom, or fixed hoist.
- ☐ Place chairs on tracks.
- ☐ Place collection ports lower to use gravity.
- ☐ Place ramps over thresholds.
- ☐ Place rollers at edge of table.
- ☐ Place shelf carts or shelves on elevators.
- ☐ Use a small electric portable lifting device.
- ☐ Use a conveyor belt.
- ☐ Use a forklift.
- ☐ Use a lever type (fishing pole) device.
- ☐ Use a rolling lever or portable rollers.
- ☐ Use a two-pole suspension system (stretcher).
- ☐ Use a vacuum tube conveyor system.
- ☐ Use an overhead suction lifting device.
- ☐ Use an overhead tube fed dispenser system.
- ☐ Use lighter trays or bins.
- ☐ Use lubricants to reduce friction.
- ☐ Use slip sheeting.
- ☐ Use special jack attachment.
- ☐ Use customized/redesigned hand trucks.
- ☐ Use special applications dolly.

Assistance To Hold in Place

- ☐ Add third hand or document holder.
- ☐ Add weight to lower part of item.
- ☐ Install a vice.
- ☐ Install special attachment for jacks.
- ☐ Place item on overhead swinging arm.
- ☐ Use a customized or stock jig.
- ☐ Use a hand tool.
- ☐ Use overhead hoist, pulleys, or arms.
- ☐ Use scaffolding or overhead gantry.
- ☐ Anchor tool to resist its force.
- ☐ Install tool balancer.
- ☐ Provide a guide attached to tool to hold work.

Decrease Repetitions

- ☐ Use a hand tool for part of process.
- ☐ Partially automate the process.

- ☐ Completely automate the process.
- ☐ Use a bigger tool.

Provide Worker Support

- ☐ Provide a chair.
- ☐ Provide adjustable chairs.
- ☐ Install a bar or step stool under workstation.
- ☐ Place a chair on gantry for extended work.
- ☐ Provide adjustable footrest.
- ☐ Provide adjustable armrests.
- ☐ Place recess in tables to support arms.
- ☐ Bevel all workstation edges and railings.
- ☐ Provide a chest support for forward leaning.
- ☐ Provide a lumbar roll or supportive cushion.
- ☐ Attach adjustable armrests to work surface.
- ☐ Provide alternate seating devices.
- ☐ Provide sit/stand chair.

Facilitate Access

- ☐ Angle bins toward the worker.
- ☐ Alter doorway/porthole to decrease awkward posturing.
- ☐ Modify trays on conveyor devices.
- ☐ Interchange shelves, bins, trays, and drawers.
- ☐ Make controls remote.
- ☐ Use a tool or reaching device.
- ☐ Place extra rung on ladders.
- ☐ Provide step stools.

Controlling Environmental Conditions

- ☐ Allow personal control of heat or cooling.
- ☐ Provide reporting mechanism for questions and concerns.
- ☐ Provide personal heating or cooling devices.
- ☐ Check climatic monitoring systems for accuracy.
- ☐ Install insulating dividers.
- ☐ Call in HVAC consultant.
- ☐ Provide protective clothing.
- ☐ Allow variability in workers' attire.

Facilitate Information Input

- ☐ Relocate meters, visual displays, or other input devices.
- ☐ Install adjustable lighting.
- ☐ Provide headphones for audio input.
- ☐ Provide lighted magnifying glass.

- ☐ Provide glare screen or reposition monitor.
- ☐ Increase size of dials or visual display equipment.
- ☐ Allow adjustability in location and display features.
- ☐ Provide earplugs.

Facilitate Control Systems

- ☐ Make handles for different functions and shapes.
- ☐ Make control handles or buttons larger.
- ☐ Place sleeves or extenders on levers or controls.
- ☐ Provide personal choice in hand or foot control selection.
- ☐ Make foot or hand controls movable.
- ☐ Interchange buttons, levers, dials, or smooth or beveled knobs.
- ☐ Move all overhead controls down.
- ☐ Place controls inset into table.

Maintenance Programs

- ☐ Establish a process for requesting maintenance.
- ☐ Set up procedures to identify maintenance need.
- ☐ Establish regular maintenance schedule.
- ☐ Initiate equipment inspection program.
- ☐ Establish flooring inspection program.
- ☐ Initiate inspection program for tool sharpness.
- ☐ Initiate inspection program for tool accuracy and calibration.

Tool Considerations

- ☐ Change tool handle design by altering angle.
- ☐ Change handle design by adding padding.
- ☐ Change handle by altering tool shape.
- ☐ Alter tool size, diameter, weight, or length.
- ☐ Simplify tool adjustment.
- ☐ Use several fixed tools rather than one adjustable one.
- ☐ Make tools electric-, air-, or battery-operated.
- ☐ Store tools in most convenient position.
- ☐ Provide the appropriate tool.

Worksheet 6–1 Templates for Reporting Ergonomic Analysis Results: Instructions and Procedures

The purpose of this step is to provide a consistent format for reporting the results of an ergonomic analysis. The long-term nature of any ergonomic process requires that available information be as comprehensive as possible. The templates will help to provide a structure for analyses.

PROCEDURES

1. Based upon initial interviews and subsequent contact with the subject of the analysis, select from the enclosed templates those reports most likely to help support your suggestions for positive change in the work environment.
2. Make copies of the necessary forms and fill in using information from preceding worksheets.

Company Name

Ergonomic Analysis of
Job Title

Address/Location of Analysis

Conducted By

Your Name

Month/Year of Analysis

Your Company Name

INTRODUCTION TEMPLATE

On	*(Date)*
An ergonomic analysis was performed at located at	*(Company name)*
	(Address)
The major product produced at this facility is	
Information contained in this report was supplied by	
Plans for the future include	
They are interested in	*(What they would like to happen after implementing suggestions in this report)*

They would also like | *(Any considerations or secondary concerns)*

The purpose of this report is

(Company name) employs *(no.)* persons working *(no.)* shifts.

In the past three years | *(Any changes or programs that occurred in the past three years)*

BACKGROUND INFORMATION TEMPLATES

The subject of this ergonomic analysis was ____(Job title)____

____(no.)____ persons are employed to perform this job function.

____(Job title)____ are responsible for

(Narrative job description)

This position requires ____(no.)____ days work per week.

Employees are given ____(no.)____ ____(paid/unpaid)____ minute coffee breaks and

____(no.)____ ____(paid/unpaid)____ minute meal breaks.

The total job exposure* for this job is ____(no.)____ hours per day.

*Amount of time that this person spends performing work.

DUTY LIST REPORT TEMPLATE

While performing the job of | *(Job title)*

the worker performs the following duties:

(Duty description)

This is performed | *(no.)* | hours per day.

(Duty description)

This is performed | *(no.)* | hours per day.

(Duty description)

This is performed | *(no.)* | hours per day.

(Duty description)

This is performed | *(no.)* | hours per day.

DUTY/TASK BREAKDOWN REPORT TEMPLATE

While performing the duty of | *(Job duty)* |
| |

the worker performs the following tasks:

| *(Task description)* |
| |
| *(Task description)* |
| |
| *(Task description)* |
| |
| *(Task description)* |
| |
| *(Task description)* |
| |
| *(Task description)* |
| |
| *(Task description)* |
| |

TASK QUANTIFICATION REPORT TEMPLATE

The task of *(Task description)*

which takes *(no.)* minutes to complete,

is performed at a rate of ____ per hour

for ____ hours per day

for a total of ____ times per day.

The task of *(Task description)*

which takes *(no.)* minutes to complete,

is performed at a rate of ____ per hour

for ____ hours per day

for a total of ____ times per day.

The task of *(Task description)*

which takes *(no.)* minutes to complete,

is performed at a rate of ____ per hour

for ____ hours per day

for a total of ____ times per day.

ERGONOMIC RISK FACTORS REPORT TEMPLATE

The following is a list of ergonomic risk factors identified for the job title of

An ergonomic risk factor is a condition or practice that can decrease worker comfort, act as an obstacle to maximum productivity and consistent quality, and increase the incidence of cumulative trauma disorders.

These risk factors were identified while performing the *(Job duty or task)* of

SUGGESTED RECORDING FORMAT FOR RISK FACTORS

Worker assumes an end-range position at the	(Body part)
(Position)	
(no.) times per task.	

Worker maintains an unsupported position at the	(Body part)
(Position)	
(no.) times per task.	

Worker applies a	(Type)		forceful exertion
with the	(Body part)		
to overcome	(no.)	pounds (kilograms)	
(no.) times per task.			

(Body part)	is exposed to
(Environmental condition)	
(no.) times per task (hours per day).	

Worker exhibits signs of behavior that may be an indication of excessive energy demands.
(Signs of excessive energy demands)

INTERVENTION DISCOVERY TEMPLATE

The following suggestions represent a comprehensive list of actions that may be taken to provide a more effective work environment. Implementation of these interventions will help to increase worker comfort level, facilitate maximum productivity and quality, and reduce the incidence of many cumulative trauma disorders.

Interventions

SUMMARY AND DISCUSSION

There is no template for a summary and recommendation section of a report. In this section, the readers should be driven to action. In many cases, this is the first section that will be read. If the information contained in this section is judged to be valuable, then the rest of the report will be considered. Use your understanding of the motivating factors in the unique corporate culture to structure this section. Often, it is a summary of the previous sections.

State need.

State solution.

Support solution.

Worksheet 7-1 Building an Ergonomic Intervention Source File: Instructions and Procedures

The purpose of this step is to compile a database of ergonomic interventions, services, and products.

PROCEDURES FOR BUILDING AN ERGONOMIC INTERVENTION RESOURCE FILE

1. Familiarize yourself with the type of product and services available.
2. Select the sources you would like to receive information from.
3. Use the sample information request letter to ask for product information.
4. When product information is received, confirm that it is the type of information you need. Make any notes in the intervention information request log summary that will help you identify this information in the future.

SAMPLE INFORMATION REQUEST LETTER

I have recently been given the responsibility for ergonomics in my facility. I will be analyzing various workstations and making recommendations for creating the safest and most effective work environments.

As part of this process, I am trying to acquire information about the various products and services available to assist me in making these recommendations. Please send current product and pricing information to me at the address above.

Thank you for your assistance.

Sincerely,

INTERVENTION INFORMATION REQUEST LOG SUMMARY

Vendor Name	Storage Location	Date Information Requested	Date Information Received

Worksheet 7-2 Intervention Comparison: Instructions and Procedures

The purpose of this worksheet is to provide a method for an initial objective comparison of similar ergonomic products or services. In addition, the vendor questionnaire may be helpful in structuring any additional contact.

PERFORMING INITIAL INTERVENTION COMPARISON USING LITERATURE

1. Review any available information and summarize features such as model numbers, vendor name, price, and installation and delivery time.
2. Add more columns or modify existing columns to suit the type of intervention.
3. Add information about other brands or providers of ergonomic products and services.
4. Pay special attention to the names, model numbers, and prices. Different vendors may sell the same product at very different prices.

PERFORMING INITIAL INTERVENTION COMPARISON BY PHONE OR IN PERSON

1. Look over the items listed in the comparison chart in this worksheet.
2. Add, modify, or delete any information as appropriate.
3. When contacting vendors by phone, inform them of your intention to purchase the most effective product or services and state that to make that assessment you will require all information discussed to be backed up in writing.
4. Confirm that the individual you speak with can answer questions regarding pricing, delivery dates, and other individuals who have used those products. (There may be an understandable reluctance to discuss pricing.)

INTERVENTION VENDOR QUESTIONNAIRE

What are the names of customers who have installed this device in a setup similar to ours?

For companies that have had success with this product, were any other interventions initiated at the same time?

For companies that have had success with this device, were any other significant changes made in workforce, production, or processes during this time?

Has there ever been an increase in injuries as a result of this product's use?

Is there an additional charge for late night or weekend installation?

Is there an additional charge for training?
What is the final cost?
How long does delivery take?
How difficult is it to install?
How long will it take to install?
How has the installation affected productivity in the past?
Can the installation be done during normal "down time" such as at night or on weekends?
How much training is necessary?
Does the vendor provide training itself or is it subcontracted?
Is support available immediately after installation?

PRODUCT COMPARISON WORKSHEET

Product Name	Model or Identification Number	Vendor Name	Price	Features	Installation and Delivery Time	Warranty

Worksheet 7–3 Intervention Implementation Plan: Instructions and Procedures

The purpose of this step is to develop a detailed plan for the implementation of an intervention.

The following general steps are suggested for planning an intervention:

1. Presenting an introduction
2. Researching and acquiring the intervention
3. Comparing the research results
4. Ordering the intervention
5. Initiating installation and orientation
6. Stating cost
7. Measuring and reporting effectiveness
8. Requesting action

Answering the questions in this worksheet will provide an excellent outline for a narrative intervention implementation plan.

INTRODUCTION

What is the suggested intervention?

What area will it affect?

How many workers will be affected?

What need is this intervention meeting?

What is the anticipated outcome?

RESEARCHING AND ACQUIRING THE INTERVENTION

Who will be responsible for researching the providers?

Can the intervention be tested on site before purchase?

How will the on-site testing be performed?

Who will contact other users to get their references?

Who else has used the intervention and what has the effectiveness been (if known)?

What was the impact on safety at other facilities (if known)?

What was the impact on productivity at other facilities (if known)?

COMPARING THE RESEARCH RESULTS

Who are the resources for this intervention?

Who will arrange for on-site testing and analysis?

Who was or will be responsible for getting bids?

Have any bids been submitted? If so, detail bids received thus far.

What was the range of bids?

Additional considerations

ORDERING THE INTERVENTION

How long will it take to get delivery?

What are the payment terms?

INSTALLATION AND ORIENTATION

How much time will be necessary to orient workers to use the new intervention?

Who will orient managers about the new intervention?

Who will arrange for any special training for the maintenance of the new interventions?

Who will be responsible for orienting new employees about this Intervention?

Who will be responsible for orienting workers to use the new intervention?

COST

What is the cost per participant?

What is the total cost?

How will this intervention take into account the projected growth of the department in terms of space as well as number of employees?

Additional cost considerations

MEASURING AND REPORTING EFFECTIVENESS

Who will be responsible for tracking the effectiveness?

What measurement will be used to track effectiveness?

How often will the effectiveness be measured?

How often will this be reported and to whom?

What type of immediate feedback collection mechanism from workers will be available?

Additional notes

REQUESTED ACTION

What type of action does the decision maker need to take?

How soon does this action need to take place?

Who can answer additional questions about this intervention?

Additional questions and comments

Worksheet 8–1 Implementation Orientation Agenda and Checklist: Instructions and Procedures

The purpose of this worksheet is to help plan the user orientation for a new ergonomic intervention. This will help to facilitate employee and line supervisors' acceptance and compliance with the intervention. When drawing up the intervention implementation checklist, consider the various steps. Add any necessary steps before you start the process.

1. Baseline data collected
2. Delivery scheduled
3. Need for maintenance criteria established
4. Orientation planned
5. Special arrangements made for illiterate or non–English-speaking workers
6. Management oriented
7. Need for maintenance communication mechanism setup
8. Intervention delivered
9. Arrangements made for absent or vacationing employees
10. Arrangements made for new hires
11. Feedback mechanism established
12. Intervention installed
13. Additional instructions printed
14. Installed intervention tested on site
15. Training time accounted for
16. Additional instructions posted
17. Orientation performed
18. Preventive maintenance scheduled

INTERVENTION ORIENTATION WORKSHEET: ORIENTATION AGENDA

What is the intervention to be implemented?
What behavior are you trying to facilitate (for example, compliance with new procedure or equipment)?
How does not complying affect workers personally? Are they in danger or at risk in any way?
How will this intervention or behavior alleviate or reduce the risk?
When will this take effect?
Why is this being done now?
Why is it being done in this area?
What is the expected outcome?
Does management expect this to affect productivity and has this been taken into consideration?

What is the mechanism for immediate feedback, complaints, or concerns?

Can this be changed back?

If this type of intervention has been tried in the past and failed, why is it being tried again?

How long will this intervention be tried?

Is additional training available?

What are the step-by-step instructions for use?

INTERVENTION IMPLEMENTATION CHECKLIST

Intervention		
Area		
Date		
	Baseline data collected	
	Delivery scheduled	
	Need for maintenance criteria established	
	Orientation planned	
	Special arrangement made for illiterate or non–English-speaking workers	
	Management oriented	
	Need for maintenance communication mechanism setup	
	Intervention delivered	
	Arrangements made for absent or vacationing employees	
	Arrangements made for new hires	
	Feedback mechanism established	
	Intervention installed	
	Additional instructions printed	
	Installed intervention tested on site	
	Training time accounted for	
	Additional instructions posted	
	Orientation performed	
	Preventive maintenance scheduled	

Worksheet 8–2 Tracking Progress and Effectiveness Report: Instructions and Procedures

The objective of this last step is to determine and report the overall effectiveness of an ergonomic intervention or project. This short report to management should accurately reflect ongoing ergonomic progress. It should answer the question, "Did it work?" as concisely as possible.

1. Enter the date, type, and baseline measurement for the criteria used to select the job or workstation for analysis.
2. Enter the dates that the background information was completed.
3. Enter the date that the risk identification was entered.
4. Briefly state the primary risk factors identified for this workstation.
5. Enter the dates that each of the intervention steps was completed.
6. Enter the changes in baseline data in the appropriate box at one week, three months, six months, nine months, one year, two years, and three years.

ERGONOMIC PROJECT TRACKING TABLE

Facility name _____

Intervention implementation completion date _____

Work area
Workstation

Statement of Ergonomic Need

Criteria Type 1	Baseline date
Baseline Measurement	

Criteria Type 2	Baseline date
Baseline Measurement	

Criteria Type 3	Baseline date
Baseline Measurement	

Background information completion dates

Date narrative statement completed	
Date duty list completed	
Date task list completed	

MAJOR RISK FACTORS IDENTIFICATION SUMMARY

Date completed	

Risk no. 1	

Risk no. 2	

Risk no. 3	

Risk no. 4	

	Completion dates
Intervention discovery	
Report presented	
Action taken	
Orientation provided	
Implementation	

Appendix G   207

CRITERIA CHANGES OVER THREE YEARS

	Date	Criteria Type 1
1 Week		
3 Months		
6 Months		
9 Months		
1 Year		
2 Years		
3 Years		

	Date	Criteria Type 2
1 Week		
3 Months		
6 Months		
9 Months		
1 Year		
2 Years		
3 Years		

	Date	Criteria Type 3
1 Week		
3 Months		
6 Months		
9 Months		
1 Year		
2 Years		
3 Years		

Index

A

Absenteeism, as measure, 6, 11
Americans with Disabilities Act (ADA), 28
Analysis process. *See* Ergonomic analysis
Arm rests, 97
Articulating arm, monitor stand, 96
Awkward range position, 38–40, 154
    nature of, 38
    recording of, 39–40

B

Background information, 13–34
    detailed measurement, 24, 142–143
    dynamic force measurement, 23, 141
    job background information, 14–15, 133
    job definition, 14, 131–132
    job description/duty list, 16–17, 19–20, 135–136
    job exposure calculation, 18
    purpose of, 13
    report, 67–68, 177
    steps in process, 13
    task breakdown, 26–31, 33, 149–151
    task cycles, 31–32, 152
    videotaping job, 25–26, 145–147
    worker impressions, 15–16, 134
    workstation measurement, 20–22, 137–139
Body position description, 36–38, 154–156

C

Chairs, 91–93
    adjustment features, 92
    necessary features, 91–93
Consultants, getting information from, 76
Controls modification, 62, 165, 172
Creativity, in intervention discovery, 54–56
Criteria Selection and Baseline Statement, 7, 121–124

D

Decision makers and ergonomic analysis, 2–4, 117–120
    format question list, 4, 118
    identification of decision makers, 117
    report for decision maker, 3
    requests of decision maker, 3
Desktop easels, 97
Desktop edge padding, 96
Detailed measurement, 24, 142–143
    example of, 24
    worksheet for, 24, 142
Document holder, 97
Duty list, 19–20, 135
    examples of duties, 19
    information for, 132
Dynamic force measurement, 23, 141
    force gauge dynamometer, 23, 141

E

Elbow, position of, 37, 155
Environmental conditions, 48–49
    adaptation to, 48
    identification of, 48, 49
    recording of, 48–49
    types of, 154
Ergonomic analysis
    background information, 13–34
    and decision makers, 2–4, 117–120
    format question list, 4, 18, 117–120
    intervention discovery, 53–64
    intervention planning, 73–82
    needs assessment, 5–12
    planning of, 1–2
    report in, 66–71

responsibilities in, 2, 113–116
risk factors, 36–52
task breakdown, 26–31
Ergonomic coordinators
role of, 109
tools of, 109–110
training of, 109
Ergonomic devices
arm rests, 97
chairs, 91–93
desktop easels, 97
desktop edge padding, 96
document holder, 97
glare screens, 93–94
installation of devices, 100–102
monitor stands, 94–96
researching devices, 99–100
standing workstations, 96
testing devices, 100
vendors, listing of, 103
worker acceptance of, 100–101
Ergonomic Need Criteria Comparison Chart, 9, 125–128
legend to, 125–126
Ergonomic program, steps in, 105–107
Ergonomics, impact of, 5
Evaluation of devices
researching devices, 99–100
testing devices, 100
worker evaluation form, 101
Excessive energy demand, 50
indications of, 50, 154
individual differences, 50
recording of, 50
Exercise
program content options, 167
program format options, 167–168
as worker-based intervention, 59, 160

F

Federal government, ergonomic information from, 77
Fingers, position of, 38, 156
Foot, position of, 36–37, 154–155
Forceful exertion, 45–48, 154
elements of, 45
identification procedure, 45
recording of, 47
risk factor potential, 46
Force gauge dynamometer, 23, 141
Format question list, 4, 18

G

Glare screens, 93-94
features of, 93

H

Hands, position of, 38, 156
Hip, position of, 37, 155
Human Factors in Ergonomics Society, 77

I

Implementation. *See* Intervention implementation; Intervention implementation plan
Incidence rate calculation, 8, 123
calculation example, 8
criteria selection/baseline statement, 7, 121–122, 124
form for, 121, 123
sources of information, 6, 121
Injury incidence rate, as measure, 6, 8, 11, 123
Input
options for, 168–169
and process-based interventions, 60, 162
Installation
of devices, 100–102
intervention implementation, 84
Intervention discovery, 53–64
creativity in, 54–56
goal of, 54
intervention discovery guide, 167–172
intervention worksheet, 63–64
problems-solving examples, 55, 56
process-based interventions, 57, 60, 162
report, 69–70, 183
worker-based interventions, 57, 160–161
workstation-based interventions, 57–58, 60–63, 162–166
Intervention implementation
checklist for, 204
installation, 84
maintenance, 85
orientation, 84, 89, 202–203
progress/effectiveness report, 85–86, 205–208
steps in, 201
Intervention implementation plan
comparing research results, 79, 196
cost, 80, 198
installation/orientation, 80, 197
introduction, 79, 194
measuring/reporting effectiveness, 80, 199
ordering intervention, 79, 197
requested action, 80, 200
research/acquisition of intervention, 79, 195
steps in, 78–81, 193–200
Intervention planning, 73–82
intervention resource file, 74–77, 185–187
purpose of, 73
Intervention resource file, 74–77, 185–187
federal government information, 77
information from consultants, 76
information request letter, 186
log summary, 187
product comparison worksheet, 75, 192
product information/comparison, 74, 189
steps in building of, 185
university programs, 76
vendor questionnaire, 190–191
Introductory statement, report, 66, 175–176

J

Job background information, 14–15, 133
form for, 133
information for, 15

Job definition, 14, 131–132
Job description/duty list, 16–17, 19–20, 135–136
  form for, 135
  and job summary, 16–17
Job exposure calculation, 18
  example of, 18
  information for, 18, 131
Job summary, 16, 17
  examples of, 17
  information in, 16, 131

K

Knee, position of, 36–37, 154–155

L

Low back, position of, 37, 155

M

Maintenance
  intervention implementation, 85
  programs, 165–166, 172
Measurements, in ergonomic analysis, 6–12
Medical department visit log, as information source, 6, 121
Monitoring program, as worker-based intervention, 59
Monitor stands, 94–96
  articulating arm, 96
  monitor mounting methods, 94
  post and arm system, 95
  readjustment of monitor, 95

N

Neck, position of, 37, 155
Needs assessment, 5–12
  criteria entries, 10
  Criteria Selection and Baseline Statement, 7, 121–124
  department comparisons, 10
  Ergonomic Need Criteria Comparison Chart, 9, 125–128
  incidence rate calculation, 8, 11, 123
  measures used, 6–7
  subjective comfort-level survey, 6–7, 10–12, 129–130

O

Occupational Safety and Health Administration (OSHA), 200
  log, 6, 121, 125
Orientation, 84, 89, 202–203
  agenda, 202–203
  guidelines for, 89
Output
  options for, 169
  and process-based interventions, 60, 162

P

Performance time, 30–31
  determination of, 30–31
Photograph, for workstation measurement, 21–22, 137–140
Planning
  for ergonomic analysis, 1–2
  intervention planning, 73–82
Post and arm system, monitor stand, 95
Problem solving. See Intervention discovery
Procedure mandate
  types of, 168
  as worker-based intervention, 59, 161
Process-based interventions, 57, 60, 162
  and input, 60, 162
  and output, 60, 162
Product comparison worksheet, 75, 192
Product information/comparison, 74, 189
Productivity, as measure, 6, 11
Progress/effectiveness report, 85–86, 205–208
  criteria changes over three years, 208
  ergonomic project tracking table, 206
  importance of, 85
  information in, 205
  risk factors identification summary, 207
Protective equipment, 59, 161
  types of, 60, 168

Q

Quality variances, as measure, 6, 11

R

Repetitions, decreasing, 61–62, 163–164, 170–171
Report
  background information, 67–68, 177
  example report templates, 174–184
  final report format, 120
  format question list, 4, 18, 117–120
  formatting of, 65–66
  intervention discovery, 69–70, 183
  introductory statement, 66, 175–176
  risk factor identification, 68–69, 181–182
  sections of, 65–66
  summary/recommendations, 70–71, 184
Research
  information about devices, 99–100
  intervention resource file, 74–77, 185–187
Responsibilities worksheet, 2, 113–116
  responsibilities, types of, 115–116
  roles and ergonomic analysis, 2, 113–116
Risk factor identification, report, 68–69, 181–182
Risk factors
  awkward range position, 38–40, 154
  body position description, 36–38, 154–156

categories of, 36, 154
environmental conditions, 48–49, 154
ergonomic risk factor, meaning of, 35
excessive energy demand, 50, 154
forceful exertion, 45–48, 154
identification of, 153
quantifying, 41–44
risk identification examples of, 157
risk identification form, 158
risk identification procedures, 50–52
Rotation program
rotation options, 168
worker-based interventions, 59, 161

S

Shoulders, position of, 37, 155
Standing workstations, features of, 96
Subjective comfort-level survey, 6–7, 10–12, 129–130
data gathering process, 129
data processing procedures, 129
example of, 130
tallying survey results, 12
Summary/recommendations, report, 70–71, 184

T

Task
definition of, 26
examples of, 27
Task breakdown, 26–31, 33, 149-151
performance time, 30–31
procedures in, 149
task description examples, 29–30
template for task description, 28
units of completion, 27
value of, 27–28
worksheet, 151
Task cycles, 31–32, 152
breakdown procedure, 149-150
determination of, 31-32
worksheet for, 33, 152
Tool options, 166, 172
Training
content options, 167
format options, 167
as worker-based intervention, 58, 160
Trauma disorder cases, as measure, 11
Turnover, as measure, 6, 11

U

University programs, ergonomics, 76
Unsupported position, 42–44, 154
example of, 42–43
recording of, 43–44

V

Vendors
phone numbers, listing of, 103
questionnaire for, 190–191
and researching devices, 99–100
Videotaping job, 25–26, 145–147
procedure in, 25–26
recording log, 25, 147
shooting procedure, 145
tips for, 145–146
videotape log, 145

W

Worker-based interventions, 57, 160–161
exercise, 59, 160
procedure mandate, 59, 161
protective equipment, 59, 161
rotation program, 59, 161
training, 58, 160
Worker impressions, 15–16, 134
form for, 134
Workstation-based interventions, 57–58, 60–63, 162–166, 169
controls modification, 62, 165, 172
decreasing repetitions, 61–62, 163–164, 170–171
environmental modifications, 62, 164–165, 171
facilitation of access, 62, 164, 171
helping to hold in place, 61, 163, 170
helping to move, 61, 163
information input facilitation, 62–63, 165, 171–172
maintenance programs, 165–166, 172
movement assistance, 169–170
supporting worker, 62, 164, 171
tool options, 166, 172
workstation adjustment, 61, 162–163
Workstation measurement, 20–22, 137–139
calculation of measures, 21, 137–139
photograph for, 21–22, 137–140
purpose of, 21
and standards, 21
summary record, 143
Wrist, position of, 38, 155